204

Anaesthesiologie und Intensivmedizin
Anaesthesiology and Intensive Care Medicine

vormals „Anaesthesiologie und Wiederbelebung"
begründet von R. Frey, F. Kern und O. Mayrhofer

K. van Ackern H. Frankenberger
E. Konecny K. Steinbereithner (Eds.)

Quantitative Anaesthesia

Low Flow and Closed Circuit

With 68 Figures and 33 Tables

Springer-Verlag
Berlin Heidelberg New York
London Paris Tokyo

Prof. Dr. med. K. van Ackern
Institut für Anästhesiologie, Medizinische Universität zu Lübeck, Ratzeburger Allee 160, D-2400 Lübeck

Prof. Dr. Dipl.-Phys. H. Frankenberger
Fachbereich Angewandte Naturwissenschaft, Fachhochschule Lübeck, Biomedizintechnik, Stephensonstraße 3, D-2400 Lübeck

Prof. Dr. E. Konecny
Geschäftsbereich Entwicklung und Konstruktion, Drägerwerk AG, Moislinger Allee 53/55, D-2400 Lübeck

Prof. Dr. med. Karl Steinbereithner
Universitäts-Klinik für Anästhesie
und Allgemeine Intensivmedizin, Experimentelle Abteilung, Spitalgasse 23, A-1090 Wien

ISBN-13: 978-3-540-50436-8 e-ISBN-13: 978-3-642-74213-2
DOI: 10.1007/978-3-642-74213-2

CIP-Kurztitelaufnahme der Deutschen Bibliothek
Quantitative anaesthesia: low flow and closed circuit/K. van Ackern ... (Eds.). -
Berlin; Heidelberg; New York; London; Paris; Tokyo: Springer 1989
(Anaesthesiologie und Intensivmedizin; 204)
ISBN 3-540-50436-2 (Berlin ...) brosch.
ISBN 0-387-50436-2 (New York ...) brosch.
NE: Ackern, Klaus van [Ed.]; GT

Typesetting and printing: Zechnersche Buchdruckerei, Speyer
Bookbinding: J. Schäffer, Grünstadt

2119/3140-543210 - Printed on acid-free paper

Preface

Quantitative anaesthesia - exact registration of the degree of anaesthesia - has long been the anaesthetist's dream. Such precision could decrease the risk for the patient. Can this dream now be realized in the foreseeable future? And can the quantification be technically verified? This workshop was organized for these questions to be discussed by a group of experts.

Schleswig-Holstein was a logical location for such a gathering. Medical and associated technical advances have socioeconomic consequences: in Schleswig-Holstein there are nowadays more jobs in medical technology than in the traditional shipbuilding industry. The encouragement of medical research and related technical innovation is thus a central component of public policy.

One result of this emphasis was the recent setting up of a study group on medical technology, one of the aims of which is to provide a forum in Schleswig-Holstein for meetings such as this. The intention is to consolidate Schleswig-Holstein's reputation as a centre for both medical-technical development and scientific discussion.

The organizers of this scientific workshop thank the Schleswig-Holstein Ministry of Economics and Transport and Drägerwerk AG for their support.

Klaus van Ackern

Preface

List of Contents

List of Contributors

J. A. Aldrete, M.D., M.S.
Department of Anesthesiology and Critical Care, Cook County Hospital, 1835 W. Harrison Street, Chicago, IL 60612, USA

Prof. Dr. med. J. O. Arndt
Abteilung für Experimentelle Anästhesie,
Universität Düsseldorf, Moorenstraße 5, D-4000 Düsseldorf

Dr. med. J. Baum
Abteilung für Anästhesie und Intensivmedizin, Krankenhaus St. Elisabeth-Stift, Große Straße 54, D-2845 Damme

E. A. Ernst, M.D.
Department of Anesthesiology, University of Alabama at Birmingham School of Medicine, Birmingham, AL 35294, USA

Prof. Dr. Dipl.-Phys. H. Frankenberger
Fachbereich Angewandte Naturwissenschaft, Fachhochschule Lübeck, Biomedizintechnik, Stephensonstraße 3, D-2400 Lübeck

Doz. Dr. phil. H. Gilly
Klinik für Anästhesiologie und Allgemeine Intensivmedizin der Universität Wien, Spitalgasse 23, A-1090 Wien

T. Oyama, M.D.
Department of Anaesthesiology, University of Hirosaki, School of Medicine, Hirosaki, Aomori-Ken, Japan

Priv.-Doz. Dr. med. H. Schwilden
Institut für Anästhesiologie, Rheinische Friedrich-Wilhelms-Universität Bonn, Sigmund-Freud-Straße 25, D-5300 Bonn

N. T. Smith, Prof. M.D.
University of California, 225 Dickinson Street, San Diego, CA 92103-1990, USA

Prof. Dr. med. K. Steinbereithner
Universitäts-Klinik für Anästhesie und Allgemeine Intensivmedizin, Experimentelle Abteilung, Spitalgasse 23, A-1090 Wien

Priv.-Doz. Dr. med. K. Taeger
Institut für Anästhesiologie der Ludwig-Maximilians-Universität,
Klinikum Großhadern, Marchioninistraße 15, D-8000 München 70

Dr. C. F. Wallroth
Anästhesiegeräte Technik, Drägerwerk AG,
Moislinger Allee 53/55, D-2800 Lübeck

D. R. Westenskow, Prof. Ph.D.
Department of Anesthesiology, University of Utah,
50 North Medical Drive, Salt Lake City, UT 84112, USA

Summary

The workshop "Quantitative Anaesthesia" in Travemünde was concerned (dealt) with a detailed inventory of today's anaesthesia, mainly inhalation anaesthesia.

The special aspect "quantitative" subdivides into two main topics:

1. Quantification of physiological processes of anaesthetized patients
2. Quantification of equipment by varying anaesthetic systems

1. Quantification of physiological processes

The first session defined quantification of physiological processes of anaesthetized patients as all possible parameters controlling changes of pharmacokinetics and pharmacodynamics of volatile anaesthetic agents as well as of oxygen uptake.

The second session predominantly dealt with the question if already today there were any advantages with quantitative anaesthesia. Anaesthetic techniques with high flow, low flow and closed systems were compared. It soon became evident that there is a great conflict between anaesthetic techniques using routinely a relatively high flow on one hand and anaesthetic techniques and dosing systems aiming at an exact balancing of the patient on the other hand.

The usual anaesthetic systems using high gas flow allow the patient to take all he needs from the offer in excess. A closed system only offers the patient what he absolutely needs. This requires a thorough measuring technique.

Pharmacokinetic and pharmacodynamic processes should be known precisely at any moment to take advantage of such a closed system. Yet, there still are substantial gaps of knowledge concerning these processes during anaesthesia. Although there are theoretical models concerning pharmacokinetics in special there are hardly any measurements in vivo to verify those.

2. Quantification of equipment

Different anaesthetic systems led to the subject of quantification of equipment. The third session was dedicated to this subject. After having taken an inventory of possible limits of today's dosing- and monitoring units it was demonstrated how highly precise dosage systems could be realized in future. Especially, the introduction of intelligent digital systems seems to be noteworthy. These highly precise dosage systems in combination with a totally closed system will allow an exact balancing of all gases and volatile anaesthetics administered to the patient as well as the actual uptake. The precise knowledge of all administered dosages offers the possibility to analyze pharmacokinetic and pharmacodynamic processes during anaesthesia, to verify existing models and to modify those, if necessary.

Balancing of dosages and uptake leads to new parameters of monitoring as for example oxygen uptake of the organism during anaesthesia.

The fourth and last session of the workshop was dedicated to modern concepts to realize a quantitative anaesthesia. Throughout the discussions and lectures the main topics were the necessary centralisation, development of intelligent priority of monitoring functions and the use of closed controlled loops.

These could be used to keep a constant level of important parameters of anaesthesia such as the inspiratory oxygen concentration or endexpiratory concentration of volatile anaesthetics. They even might be used to control depth of anaesthesia – the essential aim of any anaesthesia.

During these four sessions the workshop presented a broad overview of possibilities but also limits and problems of today's anaesthesia with volatile anaesthetics.

What is there left to conclude?

Technical development for quantitative dosage of anaesthesia and its monitoring have made great progress. But our knowledge of physiological processes during anaesthesia is still fragmentary. This development offers the precondition to examine these biological processes much closer. This would compensate for high technical expenditure of closed systems as well as all disadvantages of a slow adaption to the actual surgical situation. Only if all gaps of knowledge will be filled it will be obvious what advantages a quantitative anaesthesia will offer.

To apply this system with the utmost safety for the patient and in order to gain enough experience and knowledge it is still necessary to solve a whole series of problems.

The workshop offered an exquisite summary of the actual knowledge as exposed by the experts. This by itself would have rendered sufficient material for a workshop. But, moreover, this meeting led to valuable suggestions and definitions of problems which will have to be solved within the future. An inquiry during the workshop showed that most of the anaesthetists present were ready to choose a moderate solution namely the use of a low-flow-technique.

What Does Quantitative Anaesthesia Mean?

K. Steinbereithner

The task of providing an overview of quantitative anaesthesia puts me in a somewhat unfortunate situation in as much as I have only been involved in work on one aspect of the overall project of developing quantitative anaesthesia methods, namely the vaporizer problem. The report of this work, in which the vaporizer is considered according to the fundamental principles expounded by Frankenberger (1985) and Wallroth (1984), will be presented by Dr. Gilly. Yet being surrounded by experts in all areas, as in the saying "Prophets of the left, prophets to the right, and the child of the world in the middle," I must request that you do not view the following comments as a comprehensive review. In my address I would like to survey the field, whose growth has made it almost impossible to maintain an overview, namely three questions:

- What is quantitative anaesthesia and what arguments can be currently given for its introduction?
- Does "real" quantitative anaesthesia already exist?
- What is the outlook for the near future?

What Is Quantitative Anaesthesia?

I would like to base my definition of quantitative anaesthesia on that of Lowe and Ernst but, by extending the concept of alveolar gas, put the main emphasis on the concentration in the brain, the most important effector system. This lets me include not only inhalation anaesthetics but all other forms of anaesthesia as well. The reasons in favor of this approach are:

- The advantages of the so-called closed system (reviewed by Ernst, this volume)
- The desirability of computerized i. v. anaesthetic infusion (Shafer et al. 1987a, b; Stoeckel et al. 1982) and interactive dosing of inhalation anaesthetics (Schwilden 1987) for the purpose or providing anaesthetic agents as required
- New ideas about the design of the anaesthesia apparatus (Schreiber and Schreiber 1987; Dick et al. 1987; Baum 1987; and others)
- The considerations of cost (the most important reason).

Table 1. The use of nitrous oxide – state of the art (based on Eger and Gaskey 1986 and others)

Use N_2O
- Induction (esp. children)
- Short procedures
- Elimination of more soluble anaesthetics (end of operation)

Replace N_2O
- Severe lung disease
- Risk of gas space expansion (air embolism, pneumothorax, pneumencephalus)
- Long anaesthesia (frequently repeated short procedures)
- Pregnancy (?)
- Infection risks (problems with immune defense mechanisms?)
- Increased ICP
- (Avoidance of nausea/vomiting necessary (?))

The main cause of this last point, aside from the fact that 1 ml isoflurane vapor costs three times that of the same amount of halothane or enflurane (see the overview given by Georgi et al. 1987), would seem to be the expected or at least much-heralded retreat from the use of nitrous oxide. Table 1 gives an indication of the current situation. The use of expensive inert gases such as xenon (solubility in blood is only 30% that of N_2O; the minimum alveolar concentration (MAC) 70 vol%; Barton and Nunn 1975) or the new derivative of sevo- or isoflurane, I 653 (blood/gas coefficient, 0.424; MAC ca. 4%; recovery time after 2-h 1.2 MAC anaesthesia, 4.7 min compared to 23.2 min with isoflurane; Eger 1987; Eger and Johnson 1987a, b) may seem to be indicated as a substitute for nitrous oxide, but their use is only possible in "minimal/low flow" and closed systems.

Quantitative Anaesthesia – The Current Situation

Table 2 lists the advantages of the closed system, according to its proponents. Points one to three are quite accurate (although Lawin's group – see Scherer et al. 1987 – accords particular significance to the conversion of body heat during prolonged operations). Yet what about the continuous measurement of the parameters listed in points four and five, i.e., the postulated lung and circulation

Table 2. Advantages of closed circuit anaesthesia (Lowe and Ernst 1981; Aldrete 1984; and others)

1) Negligible pollution
2) Optimal temperature and humidity of inspired gases (conservation of water and heat)
3) Cost reduction
4) Easy and efficient monitoring of the uptake of anaesthetic vapors (and N_2O)
5) Continuous measurement of O_2 uptake, CO_2 production, alveolar and total minute ventilation etc.

Table 3. Extension of the MAC concept (Larsen 1987; Schwilden and Stoeckel 1987)

MAC_{50}	Eger et al. 1965
MAC_{95}	De Jong & Eger 1975
MAC awake	Stoelting et al. 1970
MAC EI_{50} (endotracheal intubation)	Yakaitis et al. 1977
MAC EI_{95}	Yakaitis et al. 1979
MAC BAR_{50} (blocking adrenal response) } MAC BAR_{95}	Roizen et al. 1981
(MAC additive)	Stevens et al. 1975, Teplick 1984
BAC (brain)	Wolfson et al. 1977
IAC (individual)	White 1987

Table 4. Comparison of different MAC values (Roizen et al. 1981)

	MAC_{50}	MAC_{95}	MAC EI_{50}	MAC EI_{95}	MAC BAR_{50}	MAC BAR_{95}
Halothane	1.0 MAC 0.74%	1.2 MAC	1.3 MAC ratio 1.31	1.7 MAC	1.5 MAC ratio 1.4	2.1 MAC
Enflurane	1.0 MAC 1.68%	1.1 MAC	1.4 MAC ratio 1.36	1.9 MAC	1.6 MAC ratio 1.62	2.6 MAC
Morphine, sulfate (+60% N_2O)	—	—	—	—	1.13 mg/kg ratio 1.33	1.5 mg/kg

laboratory at the head of the operating table (Lowe and Ernst 1981). Here I have to agree with Rügheimer et al. that at present this is not a realistic prospect. Aside from the impossibility of fulfilling the criterion of having a leakproof system (Rügenheimer 1985; Zbinden and Thompson 1987), there are a number of other reasons. Although very convincing MAC theory has lost none of its heuristic value, there are significant differences between the specific MAC values required for particular therapeutic purposes, even among individual anaesthetics (see Tables 3, 4). Above all, however, there is a considerable *range* between individuals (expressed as MAC_{95}/MAC_{50}), even when analgesics are used.

Considering the host of factors which influence the MAC, including the stress-catecholamine effect, the influence of acidosis, drop in blood pressure, age, and pregnancy (see Table 5), it is clear that the current models for calculating anaesthetic uptake appear to be based on numerous, unproven, and to some extent probably even incorrect assumptions (see Table 6).

A few indications of this include:

1. Fitzal et al. (1985) showed that volatile anaesthetics, without exception are taken up more quickly by children than by adults (demonstrated by the increase in the ratio for alveolar to inspiratory concentration for isoflurane). At present, one can only speculate as to the reasons for this.

Table 5. Influences on MAC (Quasha et al. 1980; Pasch 1985)

Decrease
- Hypothermia
- Hyponatremia
- Metabolic acidosis
- Hypoxemia ($paO_2 < 38$ mm Hg)
- Arterial hypotension
- Opiates
- Tranquillizers and hypnotics
- Alcohol intoxication
- Pregnancy
- Age

Increase
- Hyperthermia
- Hypernatremia
- Catecholamine release in CNS
- Chronic alcoholism
- Opiate addiction

Table 6. Modelling of anaesthetic uptake – some unverified assumptions (adapted from Zbinden et al. 1986; Schwilden and Stoeckel 1987)

- No metabolism/no diffusion
- Varying concentrations do not affect (in a predictable way) organ function(s)
- Venous concentration equals tissue concentration
- Arterial concentration equals alveolar concentration
- Blood flow to organs/compartments (respective control data being well defined) does not depend on concentration
- Partial coefficients (blood/gas, tissue/gas) remain constant, independent of concentration
- (Anaesthetic system(s) deliver(s) well-defined concentration)

2. From the histograms of the blood gas coefficients for halothane and methoxyflurane presented by Lowe and Ernst (1981), one may infer that there is significant variability in the rate of uptake among adults as well.
3. The difference which any further changes in cardiac output and perfusion may make can be seen from the comments of the same authors, according to whom changes in blood gas coefficients and cardiac output may either cancel each other or potentiate their effects.
4. There are important differences in the effect of the various halogenated anaesthetics on myocardial function. In the data of Fitzal (1984) and Gilly et al. (1986), we find that a 98% myocardial saturation is reached within 10 min with isoflurane, whereas the saturation with enflurane (90%) or halothane (80%) lags behind at this point, and full saturation is still not reached after 40 min.
5. As a result, contractility (LV dp/dt max) hardly decreases after 10 min with isoflurane, reaching a sort of equilibrium. The negative inotropic effect of the

other two anaesthetics, in contrast, increases with further saturation, leading to a further slowing of the rate of uptake.

Thus, it is no wonder that model calculations for anaesthetic uptake can produce substantial mistakes. Lauven et al. (1987) and Westenskow et al. (1983) report, for example, a 28% variation in the uptake rates for enflurane, which necessitates a corresponding readjustment of the vaporizer settings. Zbinden et al. (1986), using a new set of calculations, have found 18%–58% deviations between in vivo and model values! Even "multiple models" using up to 18 compartments and taking the interaction among anaesthetics, hemodynamics, and respiration into account have not yielded fully satisfactory results (see, for example, Smith et al. 1972 and Mapelson 1973, reviewed by Zbinden et al. 1986). Westenskow and Loughlin and Ty Smith discuss this problem in detail in chapters in this volume.

In the search for presumed new techniques for evaluating the depth of anaesthesia and thus the pharmacodynamics and pharmacokinetics of an anaesthetic, the registration of spontaneous and evoked contractions in the lower esophagus - the muscles of which are insensitive to relaxants - has become a topic of discussion. With the exception of Evans et al., who developed the commercially available esophagus monitor LECTRON 301, none of the authors listed in Table 7 have been able to establish a constant dose-effect relationship (which implies, among other things, the danger of awareness in the paralyzed subject). There are enormous differences resulting from different surgical stimuli, too (Thomas and Atikenhead). Further, White found an AD_{95}/AD_{50} ratio of 3.97, which seems to more than triple the range of MAC values listed in Table 5. The introduction of the dichotomy contractions vs cessation of contractions as the criterion (comparable to the pain/no pain principle for analgesics) would, according to White, frequently result in a level of anaesthesia deeper than necessary.

Such problems of quantification are not restricted to inhalation anaesthetics, as Table 8 shows with respect to Cp_{50} for alfentanil. In addition to the fluctuations dependent on the phases of the operation, it is necessary to take the variation in pharmacokinetics and pharmacodynamics into consideration. All of the relevant computer programs, including very recent ones, assume average values for the various parameters (see Alvis et al. 1986; Shafer et al. 1987a, b). In the

Table 7. The lower oesophageal contractility (LOC) concept (Evans et al. 1984; Cox and White 1986; White 1987; Erickson et al. 1987; Aitkenhead et al. 1987+ Isaac and Rosen 1987)

Spontaneous (SLOC)
Frequency of contractions
Peak amplitude
Mean amplitude
Provoked (PLOC)
Peak amplitude
Oesophageal contractility index
(combining SLOC + PLOC)

Table 8. Cp_{50} of plasma alfentanil concentration ($n=37$) (Ausems et al. 1986)

Event	$Cp_{50} \pm SE$ [ng/ml]
Intubation	475±28
Skin incision	279±20
Skin closure	150±23
Spontaneous ventilation	223±13

fentanyl study recently published by Stanski's group, the plasma values expected on the basis of four relevant model algorithms diverged between −33% and +10% from actual values. Thus Ausems et al. (1986) and the Bonn group are correct when they say that one must constantly remain aware that the dose-effect relationships are relatively unreliable and therefore try to achieve values within the so-called therapeutic window instead of precisely defined concentrations.

In view of the problems described here, it is understandable that attempts to employ neurophysiological CNS effects to arrive at quantified statement (such as the principle of brain anaesthetic concentration, Wolfson et al. 1972) have found renewed and widespread interest, particularly given the impressive progress in technology and microelectronics in recent years (Litscher et al. 1987) (Table 9). This is especially the case for the various types of evoked potentials. The 40 Hz steady state evoked potential (SSEP) described by Hogan represents only the latest fashion. Unfortunately, the experience to date is not very encouraging. Apparently, neither spectral nor compressed EEG frequency and amplitude analysis (Schüttler et al. 1987; Baer et al. 1987) enables the anaesthesiologist to make fine distinctions, to say nothing of the problem posed by differences between the various i.v. and inhalation anaesthetics, but provide only indications for recognizing stages of anaesthesia that are either too deep (burst suppression) or too shallow (see Stanski et al. 1987; Bührer et al. 1987). The experience using evoked potentials is not much more positive. Anaesthesia appears to have little effect on VEPs and none at all on AEPs (Thurner et al. 1987). Only the SEPs are affected, and they by no means consistently (Thiel et al. 1987).

Table 9. Monitoring of anaesthetic concentrations in brain

Respiration (spontaneous)
Motor response (?) (Wolfson et al. 1977)
Endtidal concentration (Zbinden et al. 1986)
EEG (Stanski et al. 1985; Schüttler et la. 1987)
Evoked potentials (McPherson et al. 1985; Kochs et al. 1987; Thurner et al. 1987)
Somato sensory evoked potentials (SEP)
Auditory (brainstem) evoked potentials (AEP, ABEP)
40 Hz SSEP (Hogan 1987)
Visually evoked potentials (VEP)

Pasch's laconic observation, made in 1985, that "opinions as to the clinical significance of the changes in electrical activity of the brain elicited by inhalation [and here one could add: "and all other"] anaesthetics are not unanimous" still appears to be completely valid.

Prospects for the Near Future

What will the future bring? Without doubt, the dosing formulas for both i.v. and inhalation anaesthetics are useful and mathematically tractable. An example is the well-known bolus-elimination transfer (BET) scheme of Stoeckel et al. (1982). The initial bolus is supposed to fill the central distribution space, and thereafter drugs are added only to replace the amounts which are lost via elimination or transfer. However, there are still numerous unknowns or uncertainties including biological differences which are not easy to describe or quantify, to say nothing of the imperfections of quantitative monitoring - all of which will be thoroughly considered at this conference.

I would not like to turn to questions about the technology and design of the so-called closed system.

Various anaesthesia (breathing) systems are available today which appear to be suitable for "almost" closed-circuit operation (Baum and Schneider 1983). However, in view of the minimal criteria for a closed system listed in Table 10, there are serious shortcomings (Baum 1987), particularly with respect to dosaging systems. I would like to add another criterion, namely a reduction in the inertia of the systems by decreasing system volume, as in the following equation:

$$\text{Time constant} = \frac{\text{Circle system volume } (+\text{FRC})}{\text{Fresh gas flow} - \text{net gas uptake}}$$

(Conway 1984; Rügheimer 1985)

In contrast to the view of many proponents of closed-circuit anaesthesia, (CCA) the buffer effect of the current systems is still considerable.

Important technological improvements are of course necessary. Feeble compromises are dangerous and indefensible. The suggestion made by Gorsky et al. is an example: in the absence of additional precautionary measures the system

Table 10. Minimal technical prerequisites of closed systems (Droh and Rothmann 1977; Spieß 1977; Wallroth 1984)

Precise delivery of oxygen by low-flow metering tubes
Precise delivery of nitrous oxide by low-flow metering tubes
Delivery of anaesthetics by calibrated precision (injection) vaporizers
Manual ventilation or
Leak-tight ventilators

should be operated intermittently in the half-closed mode, restricting fully closed operation to 75% of the intervention time in order to minimize risks.

At the 198 Scandinavian conference, Zbinden and Thomson (1987) attempted to provide a concise summary of everything that needs to be changed (Table 11). Despite many earlier objections (Sheppard and Kouchoukös 1976), the future use of artificial intelligence does not seem unrealistic.

I share the skepticism of the Basel group (Zbinden and Thomson 1987) with regard to the real clinical value of continuously determining the exact uptake of oxygen and anaesthetic gas. For other reasons, however, we will not be able to do without this.

This symposium will probably demonstrate the lack of unanimity with respect to what closed systems and CCA mean (Nunn), letting the differences between what Smith (1984) has called the "true believers" and the "dissidents" break out in the open.

Has the future already begun? According to the available data, shown in Table 12, this appears to be the case – and if it weren't, there would not be much sense in this meeting. Surprisingly, the low-flow concept is represented in most of the newer prototypes listed in Table 1, resurrecting Virtue's 1974 suggestion. Referring to the anaesthesiologist as the eminent head pilot, as Lowe and Ernst recently did (Lowe, by the way, has worked toward developing the CCA concept since 1962; Hill and Lowe 1962), raises a series of questions, including financial ones.

In addition to leak-proof technology and new system components (Calkins and Saunders 1984), the application of CCA requires, as has repeatedly been said,

Table 11. Requirements for closed circuit anaesthesia (Zbinden and Thomson 1987)

System has to be leak-proof (new tubing material, new fittings)
System must be able to measure O_2, CO_2 and inhalation anaesthetics (new reliable analysers necessary)
Automatic control of inflow of all gases (new electronic valves)
System must be "intelligent" to take over some of anaesthetist's actions (computer-assisted feedback control systems necessary)

Table 12. "Future" anaesthesia delivery systems (Brown et al. 1984 and others)

Alabama project (Spain et al. 1984)
Arizona program (Jewett 1984)
Boston system (Cooper and Newbower 1984)
Utah system (Westenskow et al. 1984)
Anaesthesia system AS 2.1 (Wallroth 1984)
Integrated (low flow) anaesthesia system (Lassborn 1987)
Gothenburg (low flow) model (Stenquist and Sonander 1987)

- Elaborate and costly monitoring
- A high degree of redundancy to insure safety (Wallroth 1984)
- Extreme accuracy (e.g., multiplexed mass spectrometry; see Severinghaus 1987)
- Extensive computational effort (Schultz and Arnell 1984)

Further, Lassborn and Engström (1987) have recently said that, along with the necessary sophisticated monitoring of the apparatus, the automatic determination of a number of patient parameters (e.g., constant alveolar concentration, preprogrammed depth of anaesthesia) may be considered desirable, if not absolutely necessary, in the near future.

In view of the horrendous efforts and costs necessary to guarantee the function and, in particular, safety of such devices, those (Rügheimer, Altschule) admonishing to caution should not be ignored.

In closing, I feed obliged to pose two important questions: With Bain and Spoerel (1984) I ask, "Is it worth it?" And with Schönbauer: "Can we succeed?"

To both these questions, I hope for, request, and demand answers.

Leopold Schönauer, my first teacher of surgery, said as early as 1955 that "The anaesthesist must be a very specially trained physician, a physiologist, a pathologist, a pharmacologist, he must be an internist, he must understand a lot from the heart, he must be a surgeon able to follow all phases of an operation, and he must also be an excellent physicist. All of this is necessary, but where can we find such a person with universal knowledge?" This list can be easily expanded to include the following important items: pain therapy, anaesthesia in all fields and of all kinds, tube feeding, chemotherapy, all facets of intensive therapy, emergeny care, and but not least, clinical pharmacology. The developments in anaesthesiology make this last question even more pressing.

References

Aitkenhead AR, Lin ES, Thomas D (1987) Relationship between lower esophageal contractility and clinical signs of light anesthesia. Anesthesiology 67:A671

Aldrete JA (1984) A practical perspective on low, minimal and closed system anesthesia. Acta Anaesthesiol Belg 34:251–226

Altschule MD (1986) Varagies of transcapillary albumin flux. Chest 90:280–281

Alvis J, Jacobs JR, Reves JG (1986) Computer-assisted infusions of drugs. Anesthesiology 65:345

Ausems ME, Hug CC Jr, Stanski DR, Burm AGL (1986) Plasma concentrations of alfentanil required to supplement nitrous oxide anesthesia for general surgery. Anesthesiology 65:362–373

Baer GA, Harmoinen A, Parviainen M, Rorarius M, Eerola R (1987) Serum-Thiopentalwerte, spontane M.-frontalis-EMG-Aktivität und komprimierte EEG-Amplituden- und Frequenzwerte bei Thiopentalinfusionsanästhesie. Anästh Intensivther Notfallmed 22:166–170

Bain JA, Spoerel WE (1984) The balancing opinion: why was closed circuit anesthesia abandoned? Circular 1:5

Barton F, Nunn JF (1975) Totally closed circuit nitrous oxide/oxygen anaesthesia. Br J Anaesth 47:350–357

Baum J (1987) Narkosesysteme. Anaesthesist 36:393–399
[Baum J: Referat auf dieser Tagung (1987b)]
Baum J, Schneider U (1983) Die Brauchbarkeit verschiedener Narkosebeatmungsgeräte für die Minimal-Flow-Anästhesie. Anästh Intensivmed 24:263–269
Brown BR Jr, Calkins JM, Saunders RJ (eds) (1984) Future anesthesia delivery systems. Davis, Philadelphia
Bührer M, Maitra PO, Ebling WF, Stanski DR (1987) Defining thiopental's steady state plasma concentration – EEG effect relationship. Anesthesiology 67:A399
Calkins JM, Saunders RJ (1984) Components of the system: future design requirements. In: Brown BR Jr, Calkins JM, Saunders RJ (eds) Future anesthesia delivery systems. Davis, Philadelphia, pp 89–97
Conway CM (1984) Closed and low flow systems. Theoretical considerations. Acta Anaesthesiol Belg 34:257–263
Cooper JB, Newbower RS (1984) The Boston anesthesia system. In: Brown BR Jr, Calkins JM, Saunders RJ (eds) Future anesthesia delivery systems. Davis, Philadelphia, pp 207–219
Cox PN, White DC (1986) Do oesophageal contractions measure depth of anaesthesia? Br J Anaesth 57:131P
de Jong RH, Eger EI II (1975) MAC expanded: AD_{50} and AD_{95} values of common inhalation anesthetics in man. Anesthesiology 42:384–389
Dick W, Eberle B, Friesdorf W (1987) Zukünftiger Arbeitsplatz des Anaesthesisten. Versuch einer Analyse. Anaesthesist 36:1–8
Droh R, Rothmann G (1977) Das geschlossene Kreissystem. Anaesthesist 26:461–466
Eger EI II (1987) Partition coefficients of I-653 in human blood, saline, and olive oil. Anesth Analg 66:971–973
Eger EI II, Gaskey NJ (1986) A review of the present status of nitrous oxide. AANA J 54(1):
Eger EI II, Johnson BH (1987a) MAC of I-653 in rats, including a test of the effect of body temperature and anesthetic duration. Anesth Analg 66:974–976
Eger EI II, Johnson BH (1987b) Rates of awakening from anesthesia with I-653, halothane, isoflurane, and sevoflurane. A test of the effect of anesthetic concentration and duration in rats. Anesth Analg 66:977–982
Eger EI, Saidman LJ, Brandstater B (1965) Minimum alveolar anesthetic concentration: a standard of anesthetic potency. Anesthesiology 26:756–763
Erickson JP, Foss J, Kuni DR (1987) A controlled trial of efficacy of lower esophageal contractility as a measure of depth of anesthesia. Anesthesiology 67:A672
Ernst EA (1985) Closed circuit anesthesia. In: List WF, Schalk HV (eds) Refresher-Kurs ZAK 1985. pp 127–137
[Ernst EA: Referat auf dieser Tagung (1987)]
Fitzal S (1984) Inhalationsanaesthetika und Myokardfunktion – Pharmakodynamische und pharmakokinetische Untersuchungen. Beiträge zur Anaesthesiologie und Intensivmedizin, vol 4. Maudrich, Vienna, pp 52–84
Fitzal S, Germann P, Gilly H, Grünwald C, Semsroth M (1985) Aufnahme und Elimination von Isofluran und Halothan bei Kindern und Erwachsenen. Anaesthesist 34:571–577
Frankenberger H (1985) Technische Probleme bei der Anwendung von Inhalationsanästhetika. In: Dick W (ed) Kombinationsanästhesie. Springer, Berlin Heidelberg New York, pp 95–106 (Klinische Anästhesiologie und Intensivtherapie, vol 29)
Georgie R, Hoffmann R, Hempel V, Bräutigam K-H (1987) Anästhesie-Kosten. Anästh Intensivmed 28:288–295
Gilly H, Fitzal S, Steinbereithner K (1986) Einflußnahme hämodynamischer Effekte volatiler Anästhetika auf die Gesamt- und Organaufnahme. In: List WF, Fitzal S, Schalk HV (eds) Aktueller Stand der klinischen Anaesthesie. Springer, Berlin Heidelberg New York, pp 251–258 (Anaesthesiologie und Intensivmedizin, vol 190)
Gorsky BH, Hall RL, Redford JE (1978) A compromise for closed system anesthesia. Anesth Analg 57:18–24
Hill DW, Lowe HJ (1962) Comparison of the concentrations of halothane in closed and semiclosed circuits during controlled ventilation. Anesthesiology 23:291
Hogan K (1987) 40 Hz steady state evoked potentials (SSEP) during isoflurane-N_2O anesthesia. Anesthesiology 67:A402

Isaac PA, Rosen M: Lower oesophageal contractions and depth of anaesthesia and awareness. ARS Meeting, Clin Res Centre Northwick Park, Harrow, Nov 13-14, 1987. Br J Anaesth (in press)

Jewett WR (1984) The Arizona program: development of a modular, interactive anesthesia delivery system. In: Brown BR Jr (Ed) Future anesthesia delivery systems. Davis, Philadelphia, pp 185-206

Kochs E, Treede RD, Roewer N, Bromm B, Schule am Esch J (1987) Überwachung des zentralen Nervensystems mit Hilfe evozierter Potentiale. In: Schwilden H, Stoeckel H (eds) Die Inhalationsnarkose: Steuerung und Überwachung. Thieme, Stuttgart, pp 151-164 (Intensivmedizin-Notfallmedizin-Anästhesiologie, vol 58)

Larsen R (ed) (1987) Anästhesie, 2nd edn. Urban and Schwarzenberg, Munich, p 137

Lassborn O, Engström G (1987) Low flow anesthesia - the medical industrial view. Acta Anaesthesiol Scand 31 [Suppl 86]:65

Lauven PM, Schwilden H, Stoeckel H (1987) Die Dosierung volatiler Anästhetika in Rückatmungssystemen (inkl. geschlossenes System, Low- and Minimal-Flow). In: Schwilden H, Stoeckel H (eds) Die Inhalationsnarkose: Steuerung und Überwachung. Thieme, Stuttgart, pp 21-31 (Intensivmedizin-Notfallmedizin-Anästhesiologie, vol 58)

Litscher G, Pfurtscheller G, Schwarz G, List W (1987) Akustisch evozierte Hirnstammpotentiale. Voraussetzungen für klinische Anwendungen, Datenqualität und Fehlerquellen. Anaesthesist 36:555-560

Lowe HJ, Ernst EA (eds) (1981) The quantitative practice of anesthesia. Use of closed circuit. Williams and Wilkins, Baltimore

Mapleson WW (1973) Circulation-time models of the uptake of inhaled anaesthetics and data for quantifying them. Br J Anaesth 45:319-334

McPherson RW, Mahla M, Johnson R, Traystman RJ (1985) Effects of enflurane, isoflurane, and nitrous oxide on somatosensory evoked potentials during fentanyl anesthesia. Anesthesiology 62:626-633

Nunn JF (1978) Closed-circuit anaesthesia. Br J Anaesth 50:733-734

Pasch T (1985) Grundlagen der Anwendung von Stickoxydul, Halothan und Enfluran - klinische Aspekte. In: Dick W (ed) Kombinationsanästhesie. Springer, Berlin Heidelberg New York, pp 13-25 (Klinische Anästhesiologie und Intensivtherapie, vol 29)

[Pellegrino E: quoted: Int Crit Care Dig 4:18 (1985)]

Quasha AL, Eger EI II, Tinker JH (1980) Determination and applications of MAC. Anesthesiology 53:315-334

Roizen MF, Horrigan RW, Frazer BM (1981) Anesthetic doses blocking adrenergic (stress) and cardiovascular responses to incision - MAC BAR. Anesthesiology 54:390-398

Rügheimer E (1985) Low-flow and closed-circuit anaesthesia. In: Dick W (ed) Kombinationsanästhesie. Springer, Berlin Heidelberg New York, pp 116-135 (Klinische Anästhesiologie und Intensivtherapie, vol 29)

Scherer R, Brendle BC, Lawin P (1987) Minimal-Flow-Anästhesie und vorgewärmte Infusionslösungen zur Vermeidung von intraoperativer Hypothermie bei Wirbelsäulenoperationen. Anästh Intensivmed 28:249-252

Schönbauer L (1955) Diskussion (Kgr Bd) Langenbecks Arch Chir 282:205

Schreiber PJ, Schreiber JM (1987) Ein elektronisches Monitoringsystem für Narkose und Narkosegerät mit integriertem prioritätsgeordnetem Alarmsystem. In: Schwilden H, Stoeckel H (ed) Die Inhalationsnarkose: Steuerung und Überwachung. Thieme, Stuttgart, pp 184-199 (Intensivmedizin-Notfallmedizin-Anästhesiologie, vol 58)

Schultz DG, Arnell WJ (1984) Getting the data: reducing confusion with the computer. In: Brown BR Jr (ed) Future anesthesia delivery systems. Davis, Philadelphia, pp 149-163

Schüttler J, Schwilden H, Stockel H (1987) EEG-Monitoring und Narkosetiefe. In: Schwilden H, Stoeckel H (eds) Die Inhalationsnarkose: Steuerung und Überwachung. Thieme, Stuttgart, pp 141-150 (Intensivmedizin-Notfallmedizin-Anästhesiologie, vol 58)

Schwilden H (1987) Die rechnergestützte interaktive Dosierung volatiler Anästhetika (AC-Prädiktor). In: Schwilden H, Stoeckel H (eds) Die Inhalationsnarkose: Steuerung und Überwachung. Thieme, Stuttgart, pp 167-173 (Intensivmedizin-Notfallmedizin-Anästhesiologie, vol 58)

Schwilden H, Stoeckel H (1987) Implikationen von Pharmakokinetik und MAC für die Dosierung. In: Schwilden H, Stoeckel H (eds) Die Inhalationsnarkose: Steuerung und Überwachung. Thieme, Stuttgart, pp 13–21 (Intensivmedizin-Notfallmedizin-Anästhesiologie, vol. 58)

Severinghaus JW (1987) Multiplexed mass spectrometry and low flow anesthesia. Acta Anaesthesiol Scand 31 [Suppl 86] 61

Shafer SL, Martinez GM, Stanski DR, Scott JC (1987a) Pharmacokinetic analysis of fentanyl administered by a computer controlled infusion pump. Anesthesiology 67: A666

Shafer SL, Siegel LC, Cooke JE, Scott JC (1987b) Performance criteria for computer controlled infusion pumps. Anesthesiolog 67: A192

Sheppard LC, Kouchoukos NT (1976) Computers as monitors. Anesthesiology 45:250–259

Smith BE (1984) The balancing opinion: dissident observations on closed circuit anesthesia. Circular 1:5–7

Smith NT, Zwart A, Beneken JEW (1972) Interaction between the circulatory effects and the uptake and distribution of halothane. Anesthesiology 37:47–58

Spain JA, Jannett TC, Ernst EA (1984) The Alabama automated closed-circuit anesthesia project In: Brown BR Jr (ed) Future anesthesia delivery systems. Davis, Philadelphia, pp 177–183

Spieß W (1977) Narkose im geschlossenen System mit kontinuierlicher inspiratorischer Sauerstoffmessung. Anaesthesist 26:503–513

Stanski DR, Hudson RJ, Homer TD, Scott JC (1985) Application of quantitative EEG power spectral analysis to anesthesia. In: Stoeckel H (ed) Quantitating, modelling and control in anaesthesia. Thieme, Stuttgart, pp 170–177

Stanski DR, Vuyk J, Ausems M, Arts R, Kramer C, Spierdijk J (1987) Can the EEG be used to monitor anesthetic depth for alfentanil with N_2O? Anesthesiology 67: A401

Stenqvist O, Sonander H-G (1987) The Gothenburg low flow model. Acta Anaesthesiol Scand 31 [Suppl 86] 63

Stevens WD, Dolan WM, Gibbons RT, White A, Eger EI II, Miller RD, de Jong RH, Elashoff RM (1975) Minimum alveolar concentration (MAC) of isofluran with and without nitrousoxide in patients of various ages. Anesthesiology 42:197–200

Stoeckel H, Schwilden H, Lauven PM, Schüttler J (1982) Prinzipien der klinischen Pharmakokinetik in der Anästhesiologie. Anästh Intensivther Notfallmed 17:3–10

Stoelting RK, Longnecker DE, Eger EI II (1970) Minimum alveolar concentration in man on awakening from methoxyflurane, halothane, ether and fluroxene anesthesia: MAC awake. Anesthesiology 33:5–9

Teplick R (1984) MAC additivity: reality or analytic error? Anesthesiology 61: A335

Thiel A, Russ W, Kafurke H, Hempelmann G (1987) Die Wirkung von Isofluran und Enfluran auf somatosensorisch evozierte Potentiale nach Stimulation des Nervus medianus. Anästh Intensivther Notfallmed 22:159–165

Thomas DI, Aitkenhead AR (1987) Relationship between lower oesophageal contractility and level of surgical stimulation. ARS Meeting, Clin Res Centre Northwick Park, Harrow, Nov 13–14, 1987. Br J Anaesth (in press)

Thurner F, Schramm J, Pasch T (1987) Wirkung von Fentanyl und Enfluran auf sensorisch evozierte Potentiale des Menschen in Flunitrazepam/N_2O-Basisnarkose. Anaesthesist 36:548–554

Virtue RW (1974) Minimal-flow nitrous oxide anesthesia. Anesthesiology 40:196–198

Wallroth CF (1984) Technical conception for an anaesthesia system with electronic metering of gases and vapours. Acta Anaesthesiol Belg 34:279–293

[Westenskow DR: Referat auf diesem Symposium]

Westenskow DR, Jordan WS (1984) The Utah system: computer-controlled anesthetic delivery. In: Brown BR Jr (Ed Future anesthesia delivery system. Davis, Philadelphia, pp 221–233

Westenskow DR, Jordan WS, Hayes JK (1983) Uptake of enflurane: a study of variability between patients. Br J Anaesth 55:598–601

Westenskow DR, Zbinden AM, Thomson DA, Kohler B (1986) Control of end-tidal halothane concentration, A: anaesthesia breathing system and feedback control of gas delivery. Br J Anaesth 58:555–562

White DC (1987) An approach to the measurement of depth of anaesthesia. ARS Meeting, Clin Res Centre Northwick Park, Harrow, Nov 13–14, 1987. Br J Anaesth (in press)
Wolfson B, Dorsch SE, Kuo T-S, Siker ES (1972) Brain anesthetic concentration – a new concept. Anesthesiology 36:176–179
Yakaitis RW, Blitt CD, Angiulo JP (1977) End-tidal halothane concentration for endotracheal intubation. Anesthesiology 47:386–388
Yakaitis RW, Blitt CD, Angiulo JP (1979) End-tidal enflurane concentration for endotracheal intubation. Anesthesiology 50:59–61
Zbinden AM, Thomson D (1987) Closed-circuit anesthesia techniques. Significance for teaching, research and the clinic. Acta Anaesthesiol Scand 31 [Suppl 86] 62
Zbinden AM, Frei F, Westenskow DR, Thomson DA (1986) Control of end-tidal halothane concentration, B: verification in dogs. Br J Anaesth 58:563–571

Oxygen Consumption During Anaesthesia

J. O. Arndt

As I sat before a blank piece of paper searching for an electrifying introduction to my lecture, a glance at the title page of the symposium program came to my aid. Depicted on the page was what, to all appearances, could have been a continuous record of the oxygen consumption of an adult human. If so this very rate of consumption – 220 ml/min – could have been that of a woman weighing 60 kg, taken pre-operatively, at an ambient temperature of 20° C, or after a heavy meal at 27° C, or that perhaps of a man in non-REM-sleep weighing 90 kg. Who knows? What I would like to stress with these considerations is that in order to use oxygen consumption as a measure of metabolism or energy expenditure, one must know the body size, that is, height and weight, the temperatures, both ambient and body, and the sex, age, as well as the state of wakefulness and the functional state of particular organs, especially the digestive tract. In short: Only the basal metabolic rate, i. e. the oxygen consumption measured under standardized or basal conditions (thermoneutral room temperature, 12 h after the last meal, muscular repose relaxed awake state) is the accepted standard for comparison [2, 6].

What concerns the metabolic effects of anaesthetics the opinion prevails that they would lower oxygen consumption. At first glance, the survey of the literature to this point in Table 1, seems to support this opinion because, at least decreases in oxygen consumptions were reported in most though not all studies. This view, however, is deceptive because of unreliable control conditions as subjects were premedicated or not, ventilated or not, and some had received succinylcholin, which increases oxygen consumption. The preanaesthetic values varied extremely widely between 193 and 294 $ml \cdot min^{-1}$, and it is obvious, that the fall in consumption during anaesthesia was proportionally greater the greater the initial value was (for details see [10]).

By the way, these measurements were performed by authorities in anaesthesiology. It is thus surprising, that such important information as body weight, sex, age, and above all, room temperature are not given, so that the causes for this strong variability can not be reconstructed.

Only the older study by Topkin and Artusio in human beings and in dogs by Brendel et al. used the accepted standard, i. e., the basal metabolic rate, as the basis for comparison, and these authors were the ones to find an increase, rather than a decrease in oxygen consumption under anaesthesia.

Table 1. Survey of the effect of anaesthetics on whole body oxygen consumption ($\dot{V}O_2$) in humans and dogs in relation to various reference values

Anesthetic	n	$\dot{V}O_2$ during Anesthesia	Change (%)	Reference Value	Relaxation	Premedication	Ventilation	Ref.
Humans								
Cyclopropane Pentothal + N_2O	23	168 $ml \cdot min^{-1}$	−22	Preinduction (224)*	—/C	+	CMV	1
Cyclopropane Ether	23	222 $ml \cdot min^{-1}$ 222 $ml \cdot min^{-1}$	+15 + 9	Day before (193)* Operation (202)*	—	—	SV	5
Halothane	6	144 $ml \cdot min^{-1}$	−13/−14	Table	—	+	SV	2
Halothane	5	169 $ml \cdot min^{-1}$	−22	Table (217)*	−/C	+	CMV	3
Methoxyflurane	20	96 $ml \cdot m^{-2} \cdot min^{-1}$	−19 −16	Table (118)* Before premedication (114)*	—	+	CMV	4
Halothane	6	161 $ml \cdot min^{-1}$	−35	Preinduction (249)*	—	+	SV	14
Droperidol/fentanyl + N_2O	6	130 $ml \cdot min^{-1}$	−32	Preinduction (223)*	C	+	CMV	14
Thiopental	8	Not reported	− 7	Preinduction	—	—	SV	15
Dogs								
Hexobarbital	12	3.7 $ml \cdot kg^{-1} \cdot min^{-1}$	+28	Preinduction (2.9)*	—	—	SV	6
Halothane 2.5%	10	119 $ml \cdot m^{-2}$	−16	Halothane 0.8% (142)*	S	—	CMV	16
Halothane 1.5%	8	4.7 $ml \cdot kg^{-1} \cdot min^{-1}$	−27	Halothane 0.2% (6.5)*	S	—	CMV	17
Halothane 1.5% + N_2O	23	4.6 $ml \cdot kg^{-1} \cdot min^{-1}$	—	None	—	+	CMV	18
Methoxyflurane 0.5% + N_2O	17	3.9 $ml \cdot kg^{-1} \cdot min^{-1}$	—	None	—	+	CMV	18
Various barbiturates	25	5.1 $ml \cdot kg^{-1} \cdot min^{-1}$	—	None	—	+	CMV	18
Ketamine + N_2O	?	7 $ml \cdot kg^{-1} \cdot min^{-1}$	—	None	—	+	CMV	18
Ether	9	7 $ml \cdot kg^{-1} \cdot min^{-1}$	+19	Preinduction (5.7)*	—	—	SV	7
Thiopental + N_2O	12	7.8 $ml \cdot kg^{-1} \cdot min^{-1}$	—	None	—	—	CMV	19
Own Data								
Thiopental	11	4.3 $ml \cdot kg^{-1} \cdot min^{-1}$	+ 8	Preinduction (3.8)*	—	—	SV	
Methohexital	11	4.7 $ml \cdot kg^{-1} \cdot min^{-1}$	+15	Preinduction (4.0)*	—	—	SV	
Etomidate	13	4.8 $ml \cdot kg^{-1} \cdot min^{-1}$	+28	Preinduction (3.4)*	—	—	SV	
Halothane 2% (insp.)	6	3.5 $ml \cdot kg^{-1} \cdot min^{-1}$	+32	Preinduction (sleep) (2.4)*	—	—	CMV	

* $\dot{V}O_2$ dimensions as in third column.
c = curare-type relaxant, s = succinylcholine, sv = spontaneous ventilation, cmv = controlled mechanical ventilation.

Moreso, since the primary effect of an anaesthetic is, after all, to induce unconsciousness, one may even raise the question as to what influences the change in vigilance alone has an oxygen consumption.

"Is it not conceivable that we should logically eliminate the question of increase in the internal muscular activity incidental to the waking condition, and consider the basal metabolism to that obtained during the post-absorptive condition, with complete muscular repose, and in deep sleep?" asked Brody already in 1915 in his treatise on natural sleep and metabolism [2].

Thus, if one wants to conduct "quantitative anaesthesia" with respect to oxygen consumption, which, by the way, is simple to measure, when using the closed system [8], then one must clarify the effects of the anaesthetic on the proposed measure of metabolism, the oxygen consumption. Since the problem at hand can not be solved on the basis of of the very incomplete literature, we have concerned ourselves in the past years with the effects of anaesthetics on the oxygen consumption in animals. I would like, therefore, to share with you some of what I have learned.

All experiments were carried out on trained dogs under basal conditions. That is, the animals were lying quietly usually on their left side following a 12 h overnight fast, and the room temperature was 24° C, which is thermoneutral for dogs. The oxygen consumption was continuously measured under spontaneous respiration using a so-called open-circuit flow-through technique according to Neuhof et al. [11], and on intubated animals using the open-circuit collection technique of Herr et al. [8].

One problem which is immediately obvious in such experiments (Fig. 1) is the influence of vigilance on oxygen consumption, which, when the animal is awake, is more than twice as high as during the subsequent non-REM sleep. The oxygen consumption normally varies strongly with the level of wakefulness being much higher in the alert than in the drowsy state, a phenomenon which is not so evident in this boxer dog (Fig. 1) whose disposition is naturally calm.

In order to show clearly the extent of these vigilance-dependent influences on oxygen consumption, we have taken the minimum (=drowsy, calm) and maximum (=alert) values of the waking state and those in non-REM sleep from numerous long-term recordings and contrasted them in the histogram shown in Fig. 2. Remember, all of these experiments were carried out under basal conditions, in fact, the minimum values corresponding to the basal metabolic rate calculated with Body's formula ($\dot{V}O_2 = 10.15 \cdot kg^{0.73}$) [6]. In spite of this, the oxygen consumption fluctuated strongly in the awake animals with strongly diverging maximum and minimum values, over the wide range of 2–8 ml O_2 $min^{-1} \cdot kg^{-1}$ but was the lowest and the least variable during natural phases of sleep, i.e., in deep non-REM sleep.

Humans show very similar variations in oxygen consumption in continuous registrations made over 24 h (Fig. 3). The subjects, by the way, were enclosed in an air-tight metabolic chamber and were allowed to move freely, but were instructed to avoid any strong muscle work. As with the animals, one finds here the values are highest and most variable during the day, while at night, the oxygen consumption reaches a minimum and varies little. The last measurements were made under basal metabolic conditions, and I would like to point out, that for subjects of normal weight, the metabolic rate is about 10% higher than in non-REM sleep, but for the heavier, and especially for the obese subjects, this difference runs to 20% to 30%.

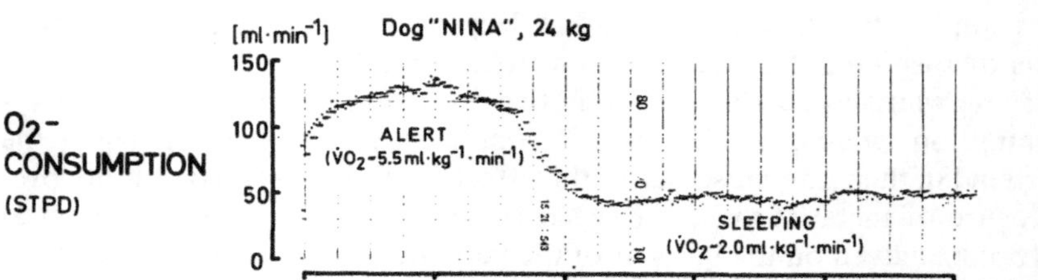

Fig. 1. Oxygen consumption ($\dot{V}O_2$) during transition from the alert resting state to natural sleep. Original recording from a dog studied under basal conditons (muscular repose, food withheld for 12 h, ambient temperature 24° C). Note that $\dot{V}O_2$ is more than halved during the transition from the alert awake state to deep sleep. (From [10])

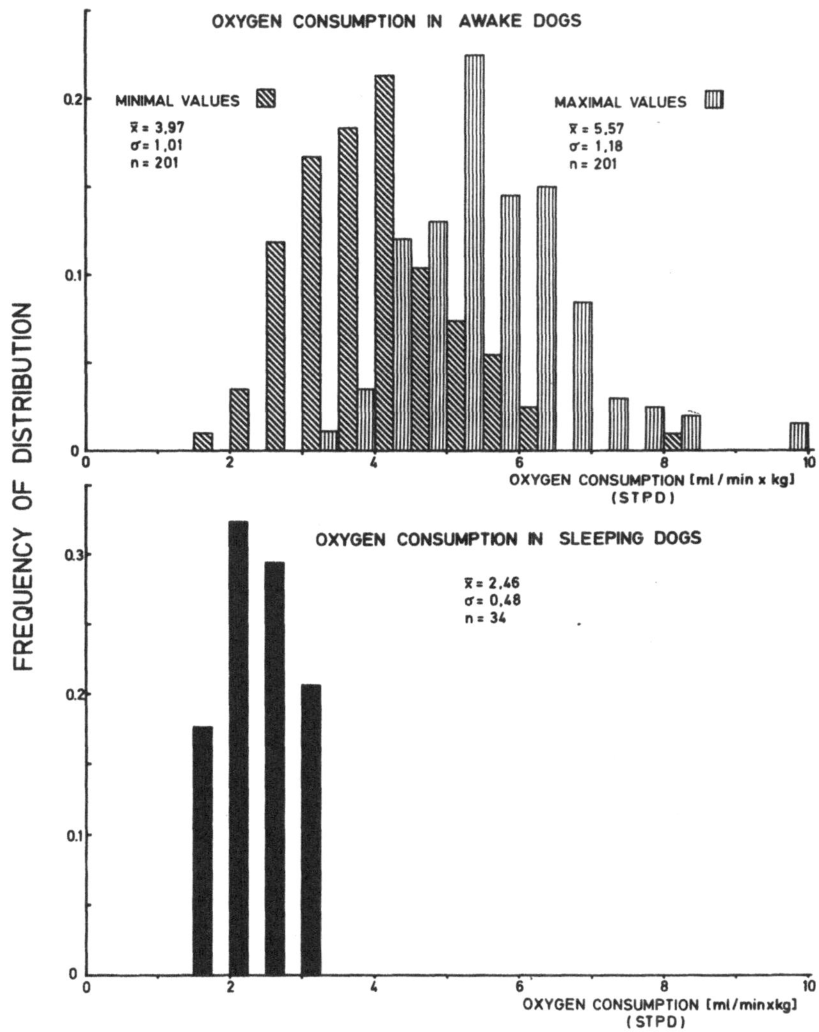

Fig. 2. Oxygen consumption at different degrees of vigilance in dogs under basal metabolic conditions. 201 experiments on 6 healthy trained dogs. The measurements were taken in the basal state (food withheld for 12 h, thermoneutral ambient temperature of 24° C, and muscular repose). Maximal values refer to the alert resting state (usually at the beginning of the experiments), the minimal values refer to the drowsy state when the animals had calmed down but were still awake, and sleep values refer to non-REM sleep.

Even under basal conditions, $\dot{V}O_2$ in the awake animals covers a wide spectrum with the maxima of the alert state being nearly twice as high as the minima of the calm state. In fact, the minimal values of 3.97 $ml \cdot min^{-1} \cdot kg^{-1}$ essentially agree with the basal values calculated by the Brody formula ($\dot{V}O_2 = 10.15 \cdot kg^{0.73}$). Oxygen consumption during sleep was lowest and fluctuated least. (From [10])

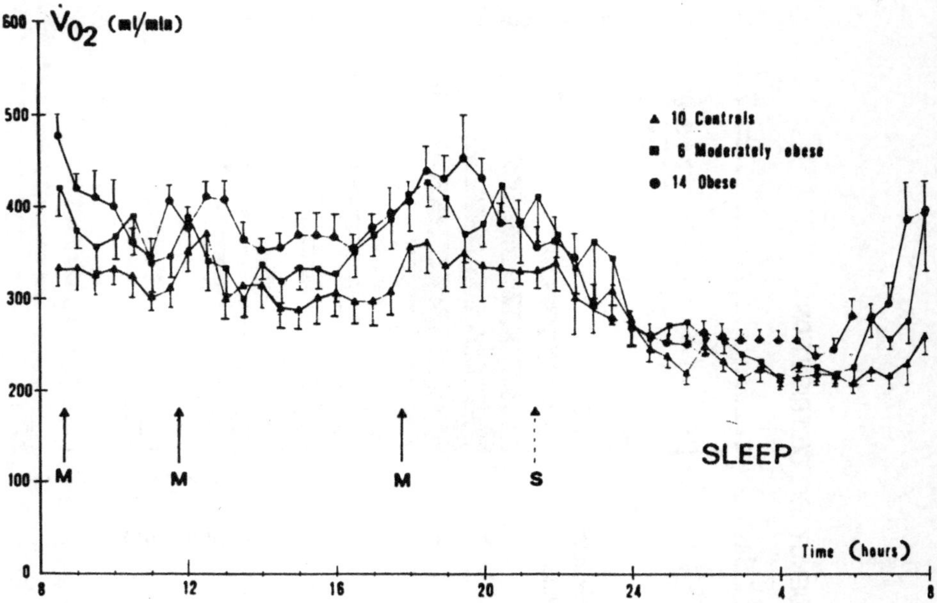

Fig. 3. 24 hours oxygen consumption in human beings. Averages (±SE) from subjects with normal and overweight. Note during natural sleep $\dot{V}O_2$ is nearly half that during day time and least variable. (From [13])

How and to what extent anaesthesia affects oxygen consumption is therefore mainly a matter of the control conditions one happens to encounter. The problem involved is apparent in Fig. 4. During anaesthesia with thiopentone, oxygen consumption is below the maxima of the alert resting state but also higher than the minimal preanaesthetic readings, which incidentally correspond to the estimated basal metabolic rate calculated by Brody's formula. Notice also the profound tachycardia. This is a typical reaction to barbiturate anaesthesia, and raises the question as to what extent the attendant increase in cadiac oxygen consumption influence the value for the total organism. Significantly, the O_2 consumption does not decrease notably when the anaesthesia is deepened by dose repetition, i.e. it is the transition from awake to anaesthesia rather than the dose of thiopentone that matters.

The effect of isoflurane anaesthesia on oxygen consumption has also been demonstrated (Fig. 5). The continuous curve gives the average oxygen consumption. The dotted horizontal lines mark the maximum and minimum resting values for these animals, as well as the values for natural sleep. Note the largest fall in oxygen consumption takes place again at the transition from the waking to the anaesthetized state at the lowest isoflurane concentration that just render the animals unconscious while further increases in concentration have no additon effect. Note also that even in deeper anaesthesia the oxygen consumption is in the neighborhood of the basal metabolic rate and in any case higher than the values measured in natural sleep.

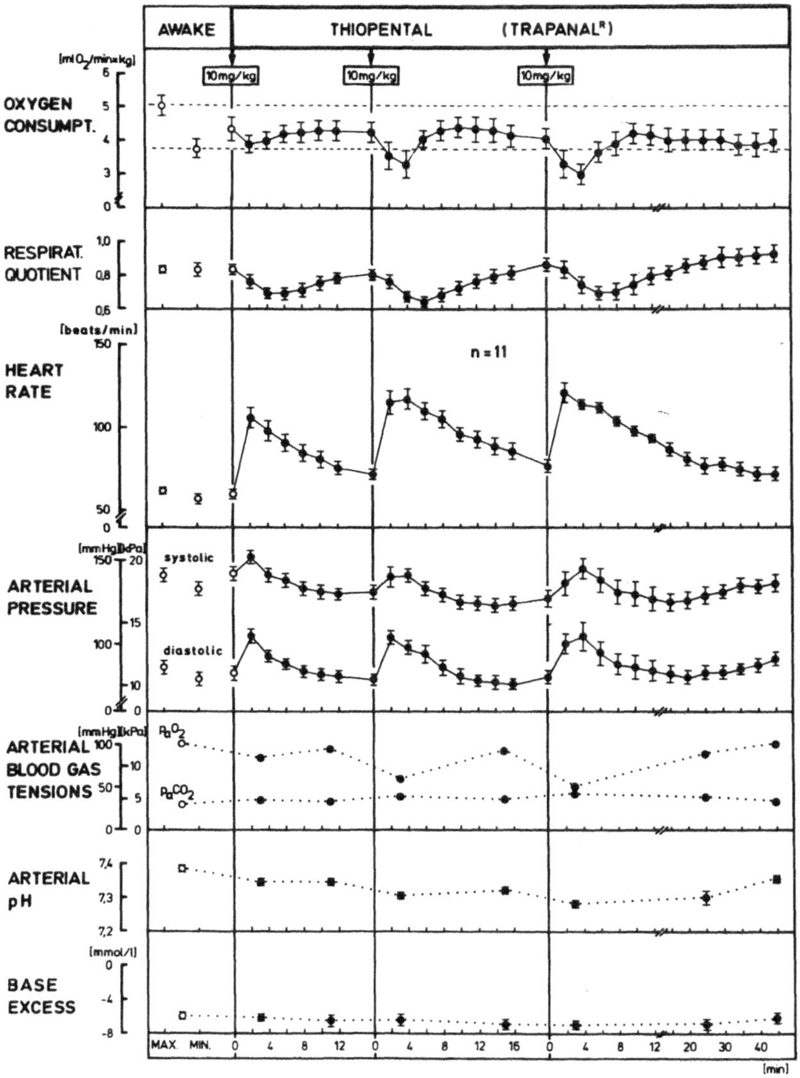

Fig. 4. Oxygen consumption during thiopentone anaesthesia in spontaneously breathing dogs. Means (±SE). Measurements were made under basal conditions, i.e. food withheld for 12 h, thermoneutral ambient temperature of 24° C, muscular repose. Dashed lines indicate maximum (awake, alert) and minimum (drowsy) preanaesthetic $\dot{V}O_2$ readings of which the latter correspond to the calculated basal metabolic rate. The repeated intravenous injection of thiopentone rendered the animals unconscious and unresponsiveness to noxious stimuli for at least 1 h.

Oxygen consumption fell below the minimum values only immediately after the injection of thiopentone because of transient apnoeas but remained between the maximum and minimum values during steady state anaesthesia when animals had resumed regular breathing. Note also that dose repetition does not affect the level of oxygen consumption. (From [10])

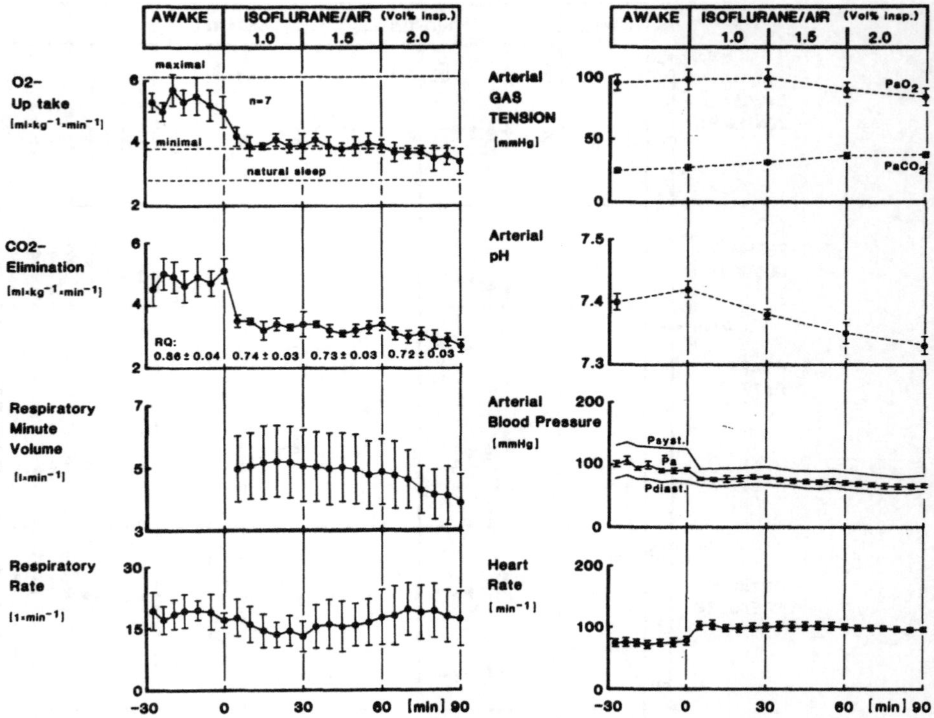

Fig. 5. The effect of isoflurane on gas exchange and various cardiorespiratory variables in dogs. Averages (± SE) from 7 experiments on 7 spontaneously breathing dogs under basal conditions (food withheld for 12 h, thermoneutral temperature of 24° C, muscular repose). The stippled lines denote the maximum values of the alert resting state, the minimal values of the drowsy resting state, and to natural sleep. Note that the decrease in O_2-uptake and CO_2-removal is strongest during the transition from wakefulness to anaesthesia while increasing concentrations have hardly any further effect which is also true with regard to the other variables. Even in deep anaesthesia oxygen consumption corresponds to the minimal or basal metabolic rate values. (From [12])

According to the findings then, it is primarily the loss of consciousness that determined the oxygen consumption in anaesthesia, and less the concentration of the anaesthetic, at least within the recommended range of concentrations.

Nevertheless, the level of oxygen consumption in anaesthesia depends on the agent used as is shown for dogs in Fig. 6. Generally, the oxygen consumption in anaesthesia is above the basal metabolic value. It is lowest for the inhalation anaesthetics, somewhat higher for the barbiturates and opiates, and highest for ketamine. Certain substance-specific side effects appear to be expressed in these differences. Barbiturates, for example, produce profound tachycardia, and skeletal muscle tone increases with opiates, and in particular ketamine.

Conclusions as to the oxygen consumption in anaesthesia depend decidedly on the reference values used. If, prior to anaesthesia, the oxygen consumption is above basal metabolic values, the accepted standard for studying metabolism, it

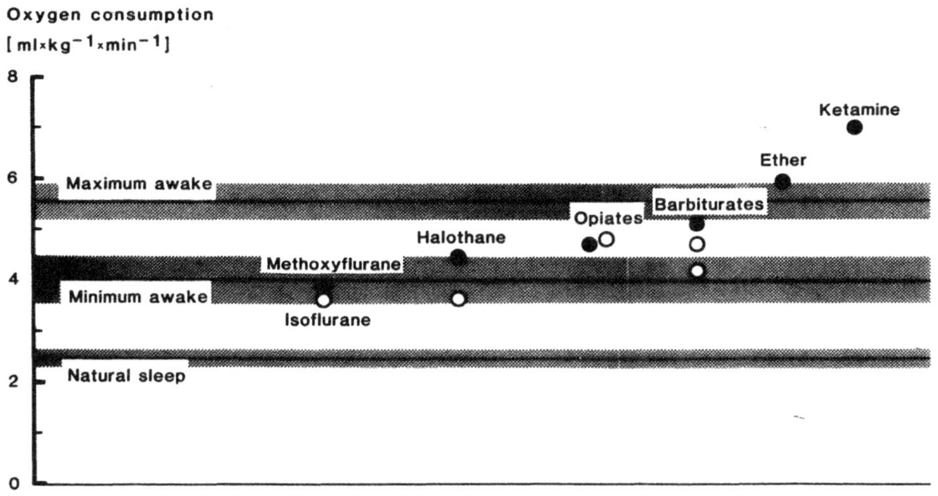

Fig. 6. Oxygen consumption during anaesthesia, natural sleep, and under basal metabolic conditions at various grades of vigilance. Survey of literature. Measurements from dogs. Horizontal lines refer to alert (maxima), drowsy state and natural sleep with shaded areas indicating standard deviations.

Oxygen consumption is lowest and fluctuates least in natural sleep (delta or non-REM sleep) and, even under basal metabolic rate conditions fluctuates strongly in relation to the state of vigilance (alert vs. drowsy). During anaesthesia oxygen consumption is above that of natural sleep is lowest and close to the metabolic rate with inhalation anaesthetics, somewhat higher with opiates and barbiturates and even above the maximum resting values with ether and ketamine. During anaesthesia, oxygen consumption in general is higher than in natural sleep and equals or exceeds the metabolic rate values depending on the agent used. (From [3, 4, 10, 12])

falls, and the higher the oxygen consumption is before anaesthesia, the more it falls in comparison with the basal metabolic rate, or what seems more reasonable to me, the minimum values found in natural sleep.

I would like to end my contribution with a short excursion into circulatory physiology which is intimately linked with the body's demand for oxygen. The close coupling between metabolism and circulation is expressed by the linear relationship between oxygen consumption and cardiac output [1, 7]. Accordingly, cardiac output can be derived by multiplying oxygen consumption by 7 and adding a species-specific constant. This formula (For dogs: Cardiac output $= 6.95 \cdot \dot{V}O_2 + 1.85$ [7]) is valid for the changes occurring with muscle work. In order to determine whether vigilance-induced changes in oxygen consumption are accompanied by similar changes in cardiac output, we implanted electromagnetic flow probes around the ascending aorta in dogs and measured oxygen consumption and cardiac output repeatedly over weeks to months.

In Fig. 7, the open symbols represent the values measured at the transition from the waking state to natural sleep. Interestingly, these data fit a line of slope 7.1, which is essentially the same as for muscle work [1]. According to this evidence then, cardiac output follows vigilance-dependent changes in oxygen con-

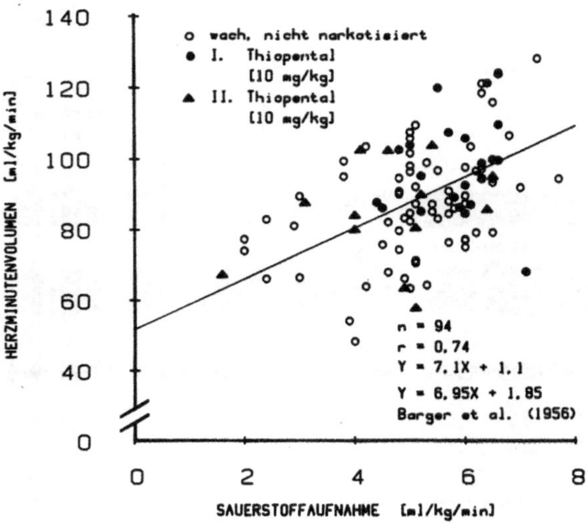

Fig. 7. Oxygen consumption ($\dot{V}O_2$) and cardiac output (CO_2) during barbiturate anaesthesia. Observations from 94 experiments on 5 trained dogs under basal conditions (muscular repose, 12 h, after the last feed, thermoneutral ambient temperature). Cardiac output was determined by chronically implanted electromagnetic flow transducer around the ascending aorta. The open symbols represent measurements before anaesthesia at various levels of wakefulness from alert via drowsy to natural sleep, the closed symbols represent measurements during anaesthesia with thiopentone which was injected twice within 5 min and had rendered the animals unconscious and unresponsive to painful stimuli for at least 30 min. Preanaesthetic values and that during thiopentone anaesthesia cluster in the same range around the regression with a slope of 7.0 (Haosheng Bi, Dissertation, University of Düsseldorf, 1983)

sumption. This hilds also for the anaesthetized state. The filled symbols in Fig. 7 represent measurements taken in a state of deep thiopentone anaesthesia lasting up to 1 h.

Due to the difficulty of measuring cardiac output in sleeping human subjects without waking them, we still do not know whether these vigilance-dependent relations between oxygen consumption and cardiac output hold for humans. With respect to Fig. 7, I would like to stress that the linear relation between these two variables, as found for dogs under barbiturate anaesthesia, applies to humans, both in neurolept and halothane anaesthesia [5]. Thus, metabolic regulation of cardiac output, that is, the adaptation of perfusion to meet the demand for oxygen is in all likelihood maintained under anaesthesia. Consequently, the oxygen consumption in anaesthesia is also a measure of cardiac output.

Conclusions

The effect of anaesthetic agents on metabolism is only then to be judged, when the oxygen comsumption is determined prior to anaesthesia under basal metabolic conditions, and perhaps even under natural unconsciousness, that is, sleep.

The claim that anaesthetic agents, in the accepted dose range, reduce the O_2 consumption is, in this generalized form, false; the oxygen consumption in anaesthesia is as high as or higher than the basal metabolic rate.

The magnitude of the oxygen consumption in anaesthesia depends upon the anaesthetic agent, insofar as the functional state of certain organs is affected by the anaesthesia. Examples of this are the high pulse rates in barbiturate anaesthesia and, above all, the increased muscle tone found in anaesthesia with ketamine, etomidate, and opiates.

The main principle of circulatory regulation, i.e. the adjustment of blood flow to meet the body's demands for oxygen, remains in effect in anaesthesia, so that the oxygen consumption may be used to make inferences about circulatory function.

The closed system allows the continuous measurement of oxygen consumption during anaesthesia and offers, therefore, an excellent opportunity to test theory in practice.

References

1. Barger AC, Richards V, Metcalfe J, Günther B (1956) Regulation of the circulation during exercise. Am J Physiol 184:613
2. Benedict FG (1915) Factors affecting basal metabolism. J Biol Chem 12:263
3. Braun U, Hensel I, Kettler D, Lohr B (1971) Der Einfluß von Methoxyflurane, Halothane, Dipiritramide, Barbiturat und Ketamine auf den Gesamtsauerstoffverbrauch des Hundes. Anaesthesist 20:269
4. Brendel W, Koppermann E, Thauer R (1954) Der respiratorische Stoffwechsel in Narkose. Pflügers Arch Ges Physiol 259:177
5. Brismar B, Hendenstierna G, Lundh R, Tokics L (1982) Oxygen uptake, plasma catecholamines and cardiac output during neuroleptnitrous oxide and halothane anaesthesia. Acta Anaesth Scand 26:541
6. Brody S (1945) Bioenergetics and growth. Waverly Press, Baltimore
7. Guyton AC, Jones CE, Coleman Th (1973) Circulatory physiology: Cardiac output and its regulation. Saunders, Philadelphia
8. Herr GP, Sullivan SF, Khambatta HJ, Crane R (1978) On-line acquisition of pulmonary gas exchange data. IEEE Trans Biomed Eng 25:83
9. Lowe HJ, Ernst EA (1981) The quantitative practice of anesthesia. Williams and Wilkins, Baltimore London
10. Mikat M, Peters J, Zindler M, Arndt JO (1984) Whole body oxygen consumption in awake, sleeping, and anesthetized dogs. Anesthesiology 60:220
11. Neuhof H, Hey D, Glaser E, Wolf H, Lasch HG (1973) Schocküberwachung durch kontinuierliche Registrierung der Sauerstoffaufnahme und anderer Parameter. Dtsch Med Wochensch 98:1227
12. Samodelov LF, Samodelov G, Arndt JO (1985) Isofluran und Gesamtsauerstoffverbrauch spontan atmender Hunde unter Grundumsatzbedingungen. Anaesthesist 34:184
13. Ravussin E, Burnand B, Schutz Y, Jequier E (1982) Twenty-four-hour energy expenditure and resting metabolic rate in obese, moderately obese, and control subjects. Am J Clin Nutr 35:566

Uptake and Distribution of Volatile Anaesthetic Substances

K. Taeger

At a symposium on the quantitative aspects of inhalation anaesthesia at which the problems of low gas flows in rebreathing systems, and in particular, closed-circle systems are emphasized, the central questions are whether, and if so, where anaesthesia concentrations should be measured, and what information these measurements give about concentrations at the site of action, the brain. A glance at the route taken by the inhalation anaesthetic, beginning at the point at which the inspiratory gas mixture is produced up to the organ central to our inquiry, the brain, reveals that we are dealing with several functional circuits: the circle system, the lungs, and the circulation. These circuits are not only connected in series, but they can also strongly influence each other (Fig. 1). Since we

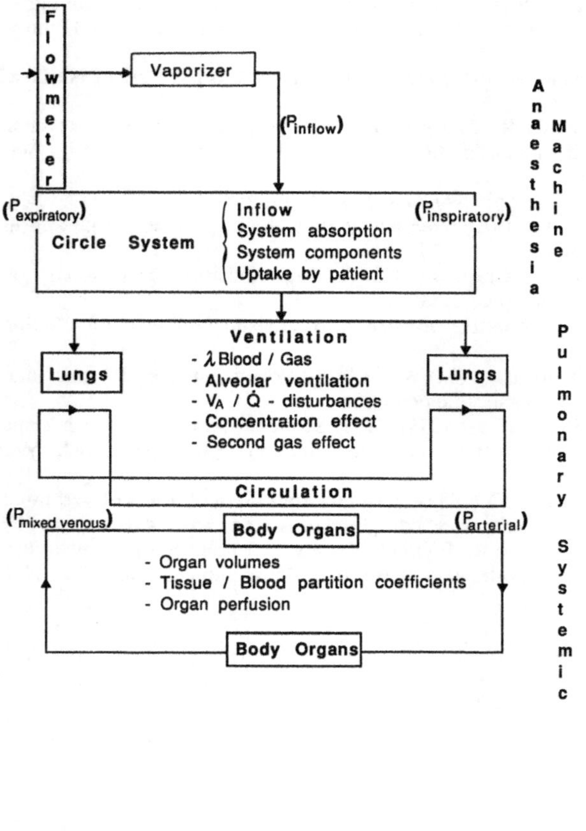

Fig. 1. Route taken by an inhalation anaesthetic. (From [4])

Table 1. Physicochemical properties of inhalation anaesthetics. (From [4])

	Halothane	Enflurane	Isoflurane	Nitrous oxide
Formula	$C_2HClBrF_3$	$C_3H_2OClF_5$	$C_3H_2OClF_5$	N_2O
Molecular weight	197.4	184.5	184.5	44.02
Vapor pressure at 20° C	242 mmHg	180 mmHg	250 mmHg	52 atm
Boiling point	50° C	56.5° C	48.5° C	−88.5° C

have to deal with various anaesthetics in this context, the physicochemical properties of these volatile anaesthetics are presented first (Table 1). The halogenated hydrocarbons halothane, enflurance, and isoflurane have boiling points between 48.5° and 56.5° C, which means that they are liquids at room temperature and must be vaporized in order to be used in anaesthesia. Since there are significant differences in the vapor pressures and boiling points of these three substances, they must be volatilized in special vaporizers which are adapted to the physicochemical properties of the particular substance. I do not wish to go into the problems here of the discrepancies between the concentrations set and delivered by vaporizers and the influence on delivered concentration of adding nitrous oxide to the gas mixture; nitrous oxide is, as its name implies, a gas at environmental temperatures.

In certain circumstances, the partial pressure of the inhalation anaesthetic in the fresh gas mixture may be significantly higher than the partial pressure required in the brain for anaesthesia. The partial pressure is reduced, step by step, as the anaesthetic passes through the many compartments, from inflow into the circle system up to the brain (Fig. 1). In rebreathing and closed systems, the fresh gas flow mixes with the expiratory gases coming from the patient. Adsorption on the components of the circle system further reduces the partial pressure of the halogenated hydrocarbons. In respiration, the gas/vapor mixture is combined with the intrapulmonary gas volume. On its way to the brain, further reductions in partial pressure occur, and are greater or smaller, depending on functional residual capacity, alveolar ventilation, intrapulmonary left-right shunts, cardiac output and its distribution within the tissues, and the solubility of the anaesthetic in blood and tissues. In order to understand the rise and fall of the inhalation anaesthetic concentrations in the various tissues it would seem important to attend to the mutual interdependence of these processes with positive and negative feedback loops. In the following, the individual processes are described independently from one another, but one must not overlook, for example, that the effects of the inhalation anaesthetic on alveolar ventilation, cardiac output, distribution of the cardiac output among the organs, autoregulation of brain perfusion, etc., may all have significant effects on the changes in partial pressure within the various compartments of the organism.

Let's consider first the changes in inhalation anaesthetic partial pressure in the aleveoli. To simplify matters, let's begin with a defined, constant concentration of the volatile anaesthetic in the fresh gas in a nonrebreathing system, where there is no mixing of fresh gases with the gas mixture in the circle system.

A measure for the speed with which the partial pressure in the alveolar space increases may be obtained by following the ratio of the end-expiratory or end-tidal partial pressure, which may be taken as representing the alveolar partial pressure, to the partial pressure of the fresh gas. The speed with which the partial pressure in the alveolar space increases depends on [2]:

- The concentration of the volatile anaesthetic in the fresh gas
- The relation between alveolar ventilation and functional residual capacity
- The solubility of the anaesthetic in the blood
- The cardiac output
- The difference in partial pressure between the alveolar gas and the mixed venous blood

It is obvious that the speed with which the partial pressure in the alveolar space increases is dependent upon the concentration of the volatile anaesthetic in the fresh gas. The higher the alveolar ventilation, the faster the alveolar partial pressure approaches that of the fresh gas. The volume of the functional residual capacity works in the opposite direction: the larger it is and the smaller the tidal volume [3], the more slowly the alveolar partial pressure of the volatile anaesthetic increases. A clinically relevant example of this would be the case of a patient with bronchial asthma. There are substance-specific differences. The effects of an increase in alveolar ventilation are most marked with ether, due to its high solubility in blood, and least marked with nitrous oxide, which is only poorly soluble. The course of the equilibration process is exponential; changes in the alveolar partial pressure are rapid at the beginning and then get slower.

We now come to the influence of the solubility of the anaesthetic in the blood upon the speed with which the partial pressure in the alveolar space increases. The solubility of an anaesthetic in blood is characterized by the blood/gas partition coefficient. This coefficient describes the relative capacity of two media, in this case blood and alveolar gas, for a given anaesthetic. For a blood/gas partition coefficient of 0.5, after equilibration, 40 parts by volume of the anaesthetic would be contained in the blood and 80 parts by volume in the alveolar gas. The larger the blood/gas partition coefficient, that is, the higher the solubility of the volatile anaesthetic in blood, the quicker the anaesthetic is taken up by the blood. A rapid uptake into the blood removes large amounts of the anaesthetic from the alveolar space and thus retards the increase in partial pressure in this space. One should not overlook the fact that the solubility of an anaesthetic in blood is by no means a constant. The composition of the blood and the changes in composition occurring during an operation have a non-significant effect on the solubility of the volatile anaesthetic in the blood. For example, solubility in blood is increased in hyperlipidemia. Thus, patients who have fasted overnight have a lower blood/gas partition coefficient than those that have not. The increased solubility in blood delays the increase in alveolar partial pressure but shortens the time required for tissue equilibration, as larger amounts of the anaesthetic are delivered to the tissues by the blood. The blood/gas partition coefficient depends also on temperature. Blood/gas and tissue/gas partition coefficients increase with decreasing temperature. Thus, a decrease in temperature delays the increase in both alveolar and tissue partial pressure. A hypothermia-induced reduction in metabolic rate, accompanied by reductions in ventilation and perfusion, results in the delay of both the increase of partial pressure in the alveolar space and the equilibration of the tissues. Thus, one would expect delays in the induction of and recovery from anaesthesia. However, since the minimal alveolar concentration (MAC) decreases in hypothermia, a reduction in body temperature has no relevant net effect.

Cardiac output also influences the increase in partial pressure in the alveolar space. The greater the cardiac output, the faster the anaesthetic is taken up into the blood, and the smaller is the increase in alveolar partial pressure. This effect is most marked for the anaesthetics which are well soluble in blood.

At the beginning of induction of anaesthesia, the tissue takes up almost all the anaesthetic that is delivered to it. This results in a large alveolar-mixed venous difference in partial pressure. With increasing tissue saturation this difference decreases; the uptake into the blood slows, less anaesthetic is taken up from the alveolar space, and the partial pressure in the alveoli can build up. The uptake of the anaesthetic by the tissues depends on the capacity and perfusion of the tissues and on the difference in partial pressure between the arterial blood and the tissues. The tissue capacity is equal to the product of the solubility of an anaesthetic in the tissue and the tissue volume. The larger the solubility or the volume, the larger is the capacity. The solubility of an anaesthetic in the tissue is described by the tissue/blood partition coefficient (Table 2). In Table 2 it may be seen that nitrous oxide not only has a low solubility in blood, but it also has a low affinity for the tissues, including fat. The halogenated hydrocarbons all have relatively similar tissue/blood partition coefficients and a comparatively high affinity for adipose tissue. The differences in tissue affinities for an anaesthetic can result in enormous differences in the capacity of identical tissue volumes. The relation of tissue capacity to blood flow determines the rate with which the partial pressure of a tissue increases. In tissues with a high capacity and a low blood flow, the concentration increases only slowly. That is, the tissue can take up anaesthetic over a very long time. In contrast, in a tissue which receives a large proportion of the cardiac output and has a limited capacity, the partial pressure increases rapidly, the tissue becomes saturated early on, and thereafter takes up no further anaesthetic. In order to understand better the significance of the individual tissues for the partial pressure difference between the alveolar space and the mixed venous blood, it is helpful, as suggested by Eger [1], to

Table 2. Partition coefficients at 37° C. (From [1])

Anaesthetic	Blood/gas	Tissue/blood						
		Brain (gray matter)	Brain (white matter)	Liver	Kidney	Muscle	Fat	Lung
N_2O	0.47	1.06		0.80		1.15	2.2	1.0
Isoflurane	1.41	2.6		2.5		4.0	45.0	
Enflurane	1.78	1.45		2.1		1.7	36.2	
Halothane	2.3	2.3	3.5	2.6	1.6	3.5	60	

Table 3. Tissue groups. (From [1])

	VRG, vessel rich group	MG	FG	VPG, vessel poor group
Relative proportion of body weight [%]	9	50	19	22
Relative proportion of cardiac output [%]	75	18.1	5.4	1.5

HVG: highly vasculated group; MG: muscle group; FG: fat group; and PVG: poorly vasculated group

group the tissues according to the proportion of the cardiac output they receive and their affinities. Eger [1] divided the various tissues of the body into the following groups: vessel rich, muscle, fat, and vessel poor (Table 3). Brain, heart, kidneys, viscera, and endocrine glands belong to the vessel rich group. Together they comprise only 9% of the body weight but receive 75% of the cardiac output. For this reason they saturate very quickly. The differences between arterial and venous partial pressures for these organs disappear correspondingly quickly. There is no significant further uptake of anaesthetic by these organs beyond 5 to 15 min after the induction of anaesthesia. The so-called muscle group is comprised of muscle and skin and makes up 50% of the body weight but, at rest, receives only one-fifth of the cardiac output. These tissues reach their distribution equilibria much more slowly, and correspondingly, their rates of uptake decrease much more slowly. The perfusion of adipose tissue, with respect to volume, is similar to that of resting muscle, but the solubility of the halogenated hydrocarbons is very high in fat. As a result, fat is able to take up volatile anaesthetics long after the end of an anaesthesia. Saturation does not occur under clinical conditions. The vessel rich group is saturated with nitrous oxide after 10 min, and with halothane after about 30 min and can not take up any more inhalation anaesthetic thereafter. While the low affinity of adipose tissue for nitrous oxide results in an almost complete saturation of the whole body after circa 200 min, halothane, thanks to the high capacity of muscle and fat, is still being taken up in significant amounts at that time. The differences among the three halogenated hydrocarbons and nitrous oxide in the time necessary to reach a 95% saturation for the individual organs are perhaps most clearly shown in the table from Zbinden ([4], Table 4). In the brain, the 95% saturation level is reached after 6.5 min for nitrous oxide, and after 10 to 15 min for halothane, enflurane, and isoflurane. In adipose tissue, this level of saturation is first reached after 7 h for nitrous oxide and not before (only after) 3 to 7 days in case of the halogenated hydrocarbons.

The concentration effect and the second gas effect [1, 2] are of some importance for the increase in alveolar partial pressure. The concentration effect describes a phenomenon occurring at high inspiratory concentrations of an inhalation anaesthetic, whereby the higher the inspiratory concentration, the faster the increase in alveolar partial pressure. Among the anaesthetics used at the moment, this applies only to nitrous oxide. The second gas effect refers to an increase in alveolar concentration of a halogenated hydrocarbon, which takes place when it is delivered simultaneously with large volumes of nitrous oxide.

Table 4. Time required for 95% saturation of the various tissue groups. (From [4])

	Halothane	Enflurane	Isoflurane	Nitrous oxide
Lungs	<1	<1	<1	<1
Kidneys	1.11	0.9	0.93	0.66
Heart	9.9	6.3	4.5	4.5
Brain	14.4	10	15	6.6
Liver	22.1	18	20.7	8.4
Musculature	207	225	240	138
Connective tissue	234	273	390	156
Adipose tissue	11250 = 7 days 19 h	5550 = 3 days 20 h	9450 = 6 days 13 h	450 = 7 h

Both effects result from a "washing out" of the alveolar space due to the increased inspiration required for the uptake of the larger gas volume, and from the increase in concentration of the gas mixture remaining in the lungs.

I would like, briefly, to address the question of disturbances of the ventilation-perfusion ratio in the lungs. An increase in the alveolar dead space has no influence on the rise in alveolar partial pressure, as long as the carbon dioxide partial pressure is kept within normal values. In contrast, intrapulmonary right-left shunts lead to differences between arterial and alveolar partial pressures. Under normal conditions, arterial and alveolar partial pressure rise in parallel. When a large shunt exists, significant discrepancies may be observed between the rises in arterial and alveolar partial pressure. The effect of shunt-like reductions in ventilation-perfusion ratios may be seen clearly for anaesthetics, such as nitrous oxide, which are poorly soluble in blood. The effect is weaker for those which are well soluble.

How closely is the cerebral partial pressure coupled to that in the alveolar space? The cerebral partial pressure follows the alveolar partial pressure with a certain delay. In hyperventilation, the alveolar partial pressure rises rapidly. However, hyperventilation causes cerebral vasoconstriction, which results in reduced perfusion of the brain. In consequence, there is a significant delay in the equilibration of partial pressures between brain tissue and alveolar gas. Hyperventilation has no significant effect on the rate of alveolar partial pressure rise for the poorly soluble alveolar partial pressure rise for the poorly soluble anaesthetics such as nitrous oxide, but the reduction in brain perfusion delays its rise in partial pressure. With hyperventilation, the alveolar partial pressure for halothane rises much more rapidly. The effects of reduced brain perfusion and higher partial pressure in the blood cancel each other almost completely so that the rise in cerebral partial pressure is not accelerated by hyperventilation. However, hyperventilation does result in a faster rise of partial pressure in the brain for the very soluble anaesthetics such as methoxyfluorane.

A low cardiac output is advantageous for the increase in alveolar partial pressure, but hinders transport to the tissues. A reduction in cardiac output by half has no significant effect on the rise in cerebral partial pressure of halothane, but causes a noticeable delay for the poorly soluble anaesthetics. If brain perfusion remains unchanged during a significant reduction in cardiac output, as for example in shock, then the more rapidly rising alveolar partial pressure is accompanied by a more rapid rise in cerebral partial pressure. As ventilation is frequently increased in shock, unexpectedly high alveolar partial pressures may develop, which, for their part, may further worsen the already impaired cardiac and circulatory functions. The raised alveolar concentration further increases the cardiovascular depression. In extreme circumstances, this positive feedback can result in cardiac arrest.

The inhalation anaesthetics themselves influence their own kinetics. The halogenated hydrocarbons are cardiodepressives and, through this action, can influence the rise of alveolar and cerebral partial pressures. Isoflurane, for example, increases the perfusion of the musculature. Since this tissue represents a very large reservoir for isoflurane, the rise in alveolar partial pressure of this substance will be delayed, because a significant alveolar-mixed venous partial pressure difference is maintained for a longer time. On the other hand, this delaying effect disappears if a massive increase in muscle perfusion occurs. In this case the musculature saturates with the speed of a vessel rich group. The metabolism of these substances can be disregarded as a factor influencing the speed with which they are taken up and distributed within the organism.

Nonrebreathing systems use large amounts of gases and anaesthetics. The problem of disposing of the expired gases also speaks against their use. For these reasons semiclosed or closed systems are most often used, which, however, further complicate the processes we have been discussing. Circle systems have a considerable volume, and contain components, for example rubber, which interact with the inhalation anaesthetics. The rates at which the inspiratory concentrations rise depend on the fresh gas inflow and the volume of the circle system, on the materials of which the circle system is made, and on the anaesthetic uptake by the patient. The gas volume of the circle system dilutes the fresh gas mixture. In order for the partial pressure in the circle system to approach that of the fresh gas, the circle system must be washed out by the fresh gas. The volume of a circle system can run up to 7 l. Low inflow rates can significantly delay the build-up of a concentration in the circle system which is necessary for anaesthesia. On the other hand, the partial pressures of the anaesthetics in the circle system may be raised quickly by using high inspiratory concentrations and high fresh gas flow initially. The adsorption of halogenated hydrocarbons on the rubber hoses and the CO_2 absorber can also substantially delay the rise of partial pressure in the circle system. Table 5, which lists the material/gas distribution coefficients at 23° C, shows that the various hydrocarbons exhibit significantly different affinities for rubber and absorber lime. Nitrous oxide is taken up by rubber in insignificant amounts. Of course, it is possible to compensate for the problem of absorption of inhalation anaesthetics by circle-system components by using high fresh gas flows and high inspiratory concentrations.

What are the consequences of closed-circle operation [2]? The gas which the patient inspires comes from two sources: the fresh gas inflow, the concentrations of which are set by the anaesthetist, and the gas which the patient has expired. As long as the lungs take up any of the anaesthetic gases, the expired gas has a lower concentration than the fresh gas. The difference in partial pressure be-

Table 5. Material/gas distribution coefficients at 23° C. (From [3])

	Halothane	Enflurane	Isoflurane	Methoxyflurane	Nitrous oxide
Conductive rubber	120	74	62	630	1.2
Polyvinyl chloride	190	120	110		
Polyethylene	26	ca 2	ca 2	118	
Absorption lime					
Fresh	1.00	1.20	0.80		
Dried	212	152	8.5		

tween the fresh gas and the gas in the circle system depends on the degree of rebreathing and the amount of the anaesthetics taken up. The smaller the inflow rate, the larger the amount of rebreathing and vice versa. The higher the alveolar concentration, the higher the concentration in the expiratory gas. If the inflow rate is large compared with the ventilation, or if the uptake of anaesthetic in the lungs is low, then the inspiratory concentrations will follow closely the fresh gas concentrations. If the ventilation is large compared with the fresh gas flow, or if large amounts of anaesthetic are being taken up, then the inspiratory concentrations will be far below that of the fresh gas. For the poorly blood soluble nitrous oxide, inspiratory and alveolar concentrations increase proportionally with increasing fresh gas inflow, as long as alveolar ventilation is kept constant. The same applies to the more blood-soluble halothane, although the differences resulting from changes in the fresh gas flow rate are greater than with nitrous oxide. At a fixed fresh gas flow of 4 l/min, an increase in alveolar ventilation results in a increase in the alveolar partial pressure of nitrous oxide. Its inspiratory concentration, however, is only minimally influenced. In the case of halothane, the alveolar partial pressure increases with increasing alveolar ventilation, but at the same time the inspiratory concentration falls significantly, limiting the rise in alveolar partial pressure. The lower the fresh gas flow, the smaller the effect of increased ventilation on alveolar concentration. In a so-called low flow system, the alveolar concentration may be relatively independent from changes in ventilation. If the system is totally closed, the anaesthetist no longer has the possibility of controlling the partial pressures in the circle system which are delivered to the patient on the inspiratory side, unless the concentrations of the anaesthetics are directly measured. By adjustment of fresh gas flow and composition, the anaesthetist controls the uptake rate. For example, if ventilation decreases, the uptake rate will only temporarily fall, because the anaesthetic begins to accumulate in the circle system. This leads to a partial pressure rise in the circle system which, in spite of the reduced ventilation, brings the patient's uptake back into line with the inflow. Thus, changes in ventilation have no long lasting effect on the partial pressures in the alveoli and brain. If the cardiac output falls, there is only a temporary decrease in rate of uptake. The reduction in uptake from the gas in the alveoli results in a rise in the partial pressure in the circle system, and this leads, in turn, to an increase in the partial pressures in the alveoli and blood. This increase could result in a further depression of cardiac function. That means, a dangerous condition could result.

If the inflow of anaesthetic into the circle system is reduced or cut off, the alveolar partial pressure will fall and the partial pressure gradient between tissue and blood, and blood and alveoli will reverse. The alveolar ventilation again determines the speed with which the partial pressure falls in the alveoli. The greater the solubility of the anaesthetic in blood and tissue is, the more slowly the concentration in the alveolar decreases. Since the degree of tissue saturation is dependent on the duration of the anaesthesia, the fall in alveolar partial pressure would be expected to be increasingly delayed with increasing duration of anaesthesia. The greater the solubility of an anaesthetic in blood and tissue is, the slower is the fall in alveolar partial pressure. As in case of the uptake of an anaesthetic, the cardiac output is of some importance for the release of the anaesthetic from the blood into the alveolar space. The higher the cardiac output, the more rapidly the anaesthetic is transported from the tissues to the lungs, and the higher the alveolar partial pressure will be. Halving of the cardiac output reduces the transport to the lungs, and the alveolar partial pressure is correspondingly lower. A low cardiac output causes a slower release of the anaesthetic from the tissues into the blood. Under these conditions the partial pressure in the brain decreases more slowly. If the reduction in cardiac output does not entail a reduction in brain perfusion, then the alveolar and cerebral partial pressure fall considerably faster than normally. The reverse is also true. If a clinically relevant decrease in pCO_2 occurs as a result of increased alveolar ventilation, the perfusion of the brain will be reduced and the transport of anaesthetics from the brain will be delayed. This is especially the case for the poorly blood-soluble anaesthetics. For the better-soluble anaesthetics, an increase in alveolar ventilation causes a significant reduction of the partial pressure in the alveoli and therefore a larger difference in partial pressure between arterial blood and the brain. The larger gradient will more or less compensate for the reduction in brain perfusion. For the poorly blood-soluble anaesthetics, hyperventilation has such a small effect on the alveolar concentration, that the decreased cerebral perfusion actually does delay the transport from the brain.

Usually, a complete equilibrium in the tissues is not reached during anaesthesia with inhalation anaesthetics. At the moment where the inflow of anaesthetic is cut off, muscles, skin, and adipose tissue are by no means fully equilibrated. For this reason, they will continue to take up anaesthetic from the blood and thereby reduce the partial pressure in the blood, at least for a time. This is especially the case for adipose tissue. The more saturated the tissues are, the less effect this uptake mechanism will have, and the longer the recovery phase will last. The circle system can influence the recovery time in a number of ways. For example, the anaesthetic dissolved in the rubber parts and absorbing lime will be set free and will limit the fall in inspiratory and alveolar concentrations. This applies mostly to the halogenated hydrocarbons. A low rate of fresh gas flow (1–3 l/min) removes the anaesthetic from the circle system only slowly and allows the rebreathing of relevant amounts of expired anaesthetic vapors. High rates of fresh gas flow wash out the circle system quickly and prevent rebreathing to a large extent. The contribution of anaesthetic released from rubber and absorbing lime is thereby also almost totally eliminated. Hyperventilation is only of limited use in shortening the time to waking. Once the vessel rich group, with its limited capacity, has more or less fully released its anaesthetics, the further course of anaesthetic release is less determined by alveolar ventilation than by the transport to the lungs from muscle and fat.

I have talked, up to now, as if the cerebral partial pressure of an inhalation anaesthetic were the sole factor determining the depth of anaesthesia. I would not like to close without at least briefly treating the aspect of individual variability in the sensitivity to inhalation anaesthetics. In one study with dogs [1], the atmospheric concentration of halothane was determined at which, in 50% of the animals, a motoric reaction to a defined pain stimulus no longer occurred. The corresponding alveolar concentration was surprisingly constant over a long period of time. There was, however, a considerable interindividual variability in the sensitivity. In another study [1], the MAC of halothane, determined on patients, varied between 0.7 and 0.85 vol%. The sensitivity, quantified as the MAC value, is influenced by a number of factors. The simultaneous administration of nitrous oxide, for example, substantially reduces the MAC of halothane. Further, there is a significant temperature dependency. The halothane-MAC decreases 5% for a 1° C drop in temperature of the organism. If the temperature increases above the normal value, the MAC value increases as well: febrile patients, for example, require more anaesthetic. There is also an age dependency. Finally, substances which cause changes in cerebral catecholamine levels influence strongly the MAC values of the halogenated hydrocarbons. For example, the MAC value may decrease by 30%, when alpha-methyl-dopa or reserpine are administered, and may increase up to 160% under the influence of *d*-amphetamine and iproniazide [1].

Thus, there is no well-defined effective concentration, known ahead of time for a particular patient, from which a "right" alveolar concentration could be pre-determined. Given this situation, the measurement of the end-expiratory concentration would seem necessary when rebreathing systems - and in particular, low-flow and closed-circuit systems - are used. However, a value, so determined, should not be interpreted as giving a reliable indication as to whether, for an individual patient, the existing concentration is effective, ineffective, or even dangerous.

References

1. Eger EI (1974) Anesthetic uptake and action. Williams and Wilkins, Baltimore
2. Mapleson WW (1984) Pharmacokinetics of inhalted anaesthetics. In: Pys-Roberts C, Hug CC (eds) Pharmacokinetics of anaesthesia. Blackwell Scientific, Oxford, pp 89-111
3. Zbinden AM (1987) Inhalationsanästhetika. Aufnahme und Verteilung. Wissenschaftliche Verlagsabteilung, Deutsche Abbott, Wiesbaden
4. Zbinden AM (1987) Inhalationsanästhetika: Pharmakokinetik. In: DAAF (Deutsche Akademie für Anästhesiologische Fortbildung), Refresher course. Aktuelles Wissen für Anästhesisten. No 13, pp 10-36

Quantitative Anaesthesia in the High-Flow System

H. Frankenberger

Inhalation Anaesthesia – System Considerations

In inhalation anaesthesia, the patient is supplied with a variable mixture of gases which have been enriched with vapors of a volatile anaesthetic. On the inspiratory side, the following media are of importance: oxygen, nitrous oxide, and one of the various volatile anaesthetics such as halothane, enflurane, or isoflurane. On the expiratory side the patient introduces carbon dioxide into the mixture.

The composition of the gas mixture is controlled by a gas and anaesthetic delivery system which is connected to the patient via a breathing circuit. These media are transported to the patient's lungs by a respirator. After diffusion through the alveoli of the lungs the dissolved O_2–N_2O mixture and the dissolved anaesthetic are carried to the various organs by the circulatory system.

The concentrations of these media must be kept within the following boundary conditions:

1. The concentration of the oxygen administered on the inspiratory side may not fall below a certain value
2. The concentration of the volatile anaesthetic administered may not rise above a certain value

The elements of this system are presented in the block diagram of Fig. 1, showing the gas and anaesthetic delivery system, ventilator system, breathing circuit, lungs (alveolar volume), circulatory system (blood volume), organs, and tissue.

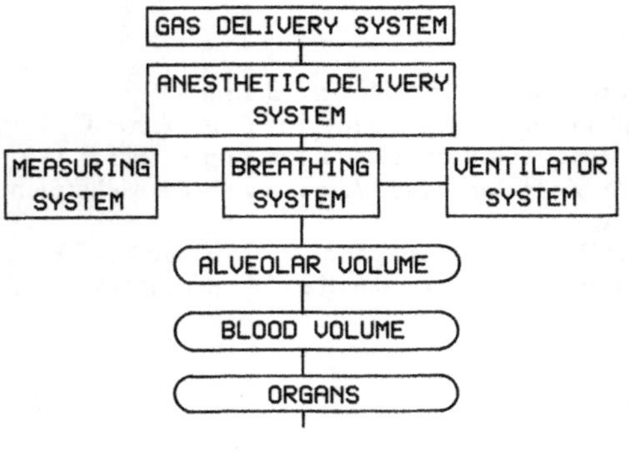

Fig. 1. Block diagram of the elements of an inhalation anaesthesia system

In order to achieve the necessary depth of anaesthesia, the patient must take up the above-mentioned gases and anaesthetics in particular amounts. Although uptake is determined by physical principles, mathematical models of these processes are, up to now, no more than approximations [1-3]. The effect of a volatile anaesthetic depends on its uptake by, distribution within, and elimination from the body. It reaches its main site of action, the central nervous system, via the circulatory system. The concentration of the anaesthetic in the blood, which is directly dependant upon its solubility in the blood, is the determining factor for the entrance of anaesthetic into the central nervous system.

From these system factors it becomes clear that a quantitative anaesthesia requires not only consideration of the technique of gas mixture and volatile anaesthetic delivery, but also consideration of the behavior of these substances at the apparatus-patient interface and their uptake by and distribution within the patient.

Classification of Anaesthesia Systems

In the majority of inhalation anaesthesias carried out today, the amount of oxygen, nitrous oxide and volatile anaesthetic delivered to the patient is in excess of the patient's oxygen consumption and anaesthetic needs. Accordingly, such anaesthesia systems may be classified as "excess systems" [4]. The excess systems include "high-flow" systems with and without CO_2 absorption as well as "low-flow" systems. Generally, high-flow systems are operated with fresh-gas flows greater than 4 l/min. Systems using fresh-gas flows of about 1 l/min are described in the literature as "low-flow" systems [5]. In contrast to high- and low-flow systems, the totally closed anaesthesia system represents a "Balanced" system. Use of a balanced system requires that only those amounts of gases and volatile anaesthetics be delivered to the patient that will be taken up by the patient in the time interval in question.

The classification scheme just described is depicted in Fig. 2. High-flow and low-flow refer to the amounts of gases administered by the delivery systems.

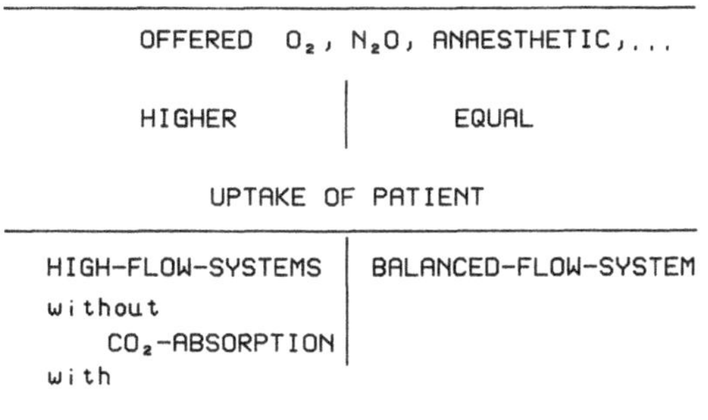

Fig. 2. Classification of anaesthesia systems

Design of High-Flow Anaesthesia Systems

Gas and Anaesthetic Delivery Systems

In Fig. 3 the fundamental elements of the delivery systems for gases and volatile anaesthetics in a high-flow anaesthesia system are depicted. Characteristically, the functional components of the delivery and breathing circuit systems are arranged in series.

When surface-evaporation vaporizers are used, the serial arrangement results in a physical coupling between the amount of gas flow and the amount of volatile anaesthetic. Within these vaporizers the gas stream is divided into two parts: one passes through the evaporation chamber and the other through a bypass (Fig. 4). The portion passing through the evaporation chamber will be fully saturated with the vapors of the liquid anaesthetic and then recombined, within the vaporizer unit, with the portion flowing through the bypass. Since the portion passing through the evaporation chamber is fully saturated with the volatile anaesthetic vapors, a fixed coupling results between the amounts of gas and anaesthetic transported. This coupling is characteristic of high-flow systems and

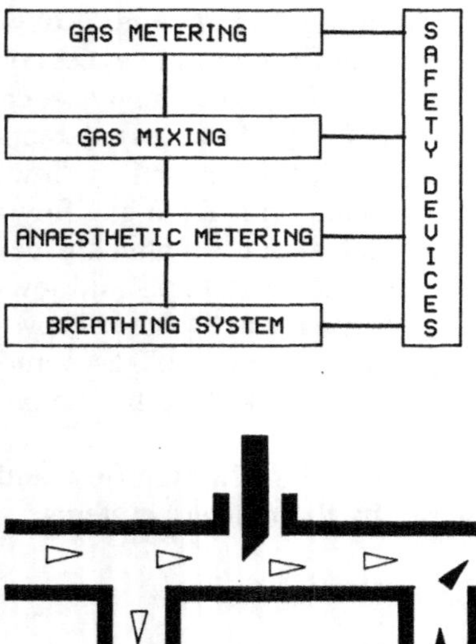

Fig. 3. Delivery system of an anaesthesia apparatus

Fig. 4. Design principle of a surface-evaporation vaporizer

does not represent a limitation for low-flow systems. It should be noted that the relation between the gas flow and the amount of anaesthetic taken up per time unit by the patient is independent of the coupling relation between amount of gas flow and the amount of volatile anaesthetic. For this reason, a different technical solution must be used with totally closed systems ("Balanced" systems).

In order to satisfy the requirements of safety standards such as DIN 13252: Inhalational anaesthetic apparatus, requirements - for safety and testing [6], high-flow systems must be fitted with several safety devices. Relevant to the inhalational anaesthesia apparatus are, for example, nitrous oxide cut offs, and a low oxygen warning signal. The Regulations on the Safety of Medical Technical Equipment requires that all anaesthesia units introduced on the market in the Federal Republic of Germany after January 1, 1988 be equipped with warning device to indicate incorrect dosages due to equipment failure [7]. This requirement may be satisfied by connecting an anaesthetic measuring device which has built-in alarm limits at the output of the anaesthetic vaporizer.

Breathing Systems

Breathing systems represent the interface between the anaesthesia apparatus and the patient. A breathing system is an assembly of components through which the patient breaths. According to Schreiber [8] breathing systems may be classified not only with respect to the proportion of rebreathing and processing of expiratory gases (CO_2 absorption), but also with respect to their design principles.

The breathing systems in Table 1, listed according to their design principles, can all be used for high-flow systems. Of these, the "To and Fro Systems" are no longer used. This is not only due to the difficulty of operating these systems, but also for functional reasons (e.g., the dead space increases with exhaustion of the CO_2 absorber and the temperature of the breathing gas can be above body temperature). The first three systems listed in Table 1 have in common that no processing of the breathing gases occurs and that the concentrations of the various gases reaching the patient from the breathing system correspond to those introduced by the gas and anaesthetic delivery systems.

The circle system, listed last in Table 1, is very widely used in the Germany. Its first implementation, worldwide, was in an anaesthesia unit developed by the Dräger company in 1925 [9] and shown in Fig. 5. While the principles of the

Table 1. Classification of breathing systems

Anaesthesia systems according to mode of use
- Systems with non-rebreathing valves
- T-piece valves
- Magill systems
- To and Fro systems
- Circle systems

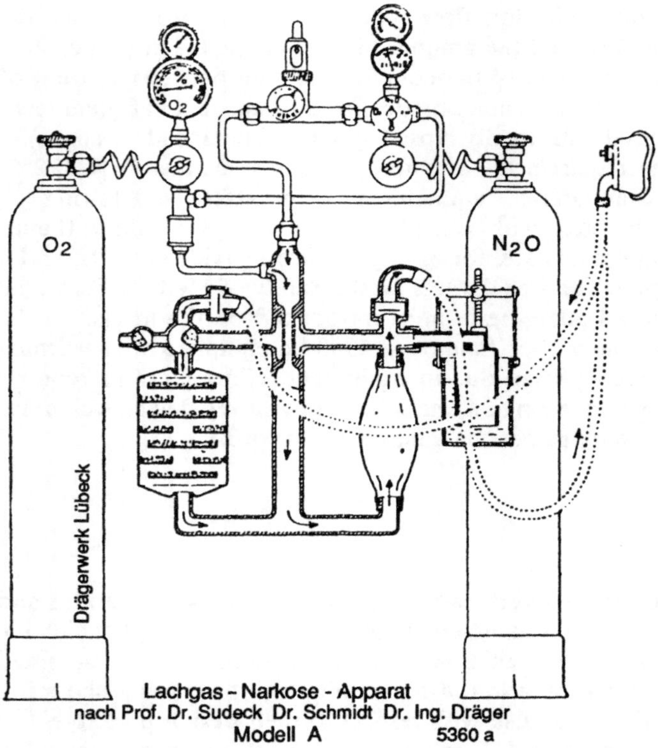

Fig. 5. Functional scheme of the first anaesthesia apparatus using the circle system: Dräger, Model A, developed in 1925 (corresponds to Dräger's figure no 40688)

system have remained unchanged since then, additions have been made allowing the connection of a respirator and measurement devices for oxygen, pressure, volume, etc.

Criteria for monitoring at the interface between patient and anaesthesia apparatus have been oriented towards preventing effects on the patient which should be avoided during anaesthesia. These effects, which can be influenced by the anaesthesia apparatus, include hypoxia, hyperoxia, alveolar collapse, barotrauma, hypocapnia, hypercapnia, anaesthesia levels which are too shallow or too deep, and so forth.

In order to avoid such effects, the specification cited above, DIN 13252, requires the use of specific monitoring devices with circle systems. These devices are used with high-flow systems, but their use is also of importance for anaesthesia systems using reduced flows of fresh gas.

Measuring devices on the inspiratory side of the breathing system may be used to quantify the parameters of the gas mixture that are administered to the patient. Those on the expiratory side may be used to measure the gas concentrations given off by the patient.

It should be noted that the amounts of gas reaching the patient from the breathing system are not necessarily identical with the amounts introduced by the gas delivery system. Deviations may be caused by leakages as well as by gas losses at the *(pop-off)* valve occurring, for example, during ventilation.

The concentrations administered from the delivery system differ from those delivered from the breathing system to the patient. The breathing system (circle system), which has a volume of circa 6 l, continually receives fresh gas, on the one side, while the respirator, on the other side, with its set parameters (frequency, volume, etc.), moves volumes of gas from the breathing system into the lungs and then from the lungs into the breathing system.

Measurement of Patient Parameters

To date, the uptake of the volatile anaesthetic in the alveolar space, its uptake and distribution in the blood, in the organs and in the tissue have been described only by very simplified linear models, as developed, for example, by Eger [1] and Lowe and Ernst [10]. These models require measurements of the inspiratory and alveolar concentrations of the volatile anaesthetic, alveolar ventilation, cardiac output, and the solubility characteristics of the anaesthetics, which differ in blood, tissue, and fat. The uptake of anaesthetic in a particular organ depends on parameters as f.i.: its size and its perfusion.

Since these values and parameters are not known for individual patients, a quantitative anaesthesia would require measurement systems which, as shown in Fig. 6, register simultaneously the necessary data for alveolar volume, circulatory system and for the various organs. Measuring instruments available to date are only of limited use towards these goals. The only parameters that can be measured satisfactorily on a continuous basis are those which are measured on the expiratory side of the breathing system, and those which are measured on a non-invasive basis such as oxygen saturation, or those parameters measured invasively within the cardiovascular system. Measurement systems which continually register the uptake and release of anaesthetic agents are not as yet in use.

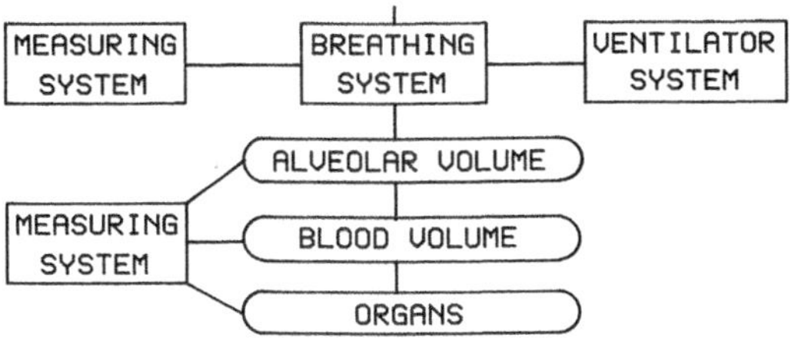

Fig. 6. Block diagram: patient parameters which may be measured

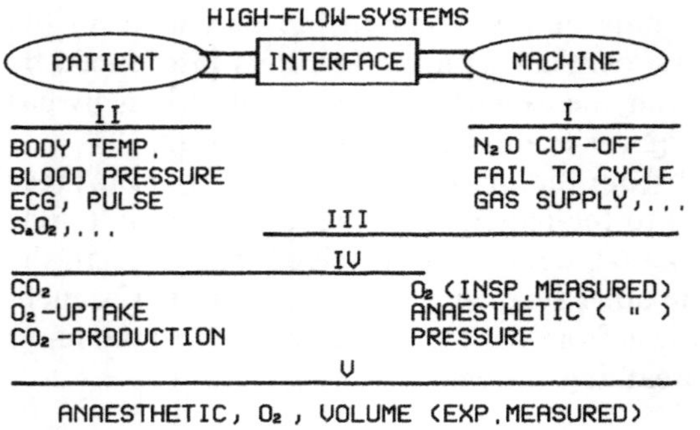

Fig. 7. Measuring and safety devices in the patient - anaesthesia apparatus system

Discussion

If one considers the anaesthesia apparatus and patient together as a system, it is possible, for "high-flow" anaesthesia systems, to quantify the amounts of gases and volatile anaesthetics delivered to the patient by the apparatus and the substances released by the patient via the lungs. At the present it is not possible to measure the uptake and distribution of these substances within the patient and only very simplified linear models exist for the representation of these processes.

During an operation and the attendant anaesthesia, changes in the uptake and distribution of the anaesthesia can take place. The application of anaesthesia in a totally closed balanced system requires that such changes be recognized early enough for the delivery system to be adjusted correspondingly. This, in turn, requires the possibility of measuring the patient parameters indicated in Fig. 6. As long as such measuring systems are not available, excess-systems in the sense of high-flow systems will remain the method of choice.

The degree to which the gas flow values exceed the amounts of uptake by the patient in a high-flow system depends principally upon the measuring systems available and upon those actually used. The aspects of the patient-apparatus system covered by the measurement and safety systems familiar today are shown in Fig. 7. If these measurement possibilities are exploited, it is possible, as long as the specified leakage rates of the customary circle systems have been verified, to reduce the gas flow rates for oxygen to approximately 1 l/min.

References

1. Eger EI (1974) Anesthetic uptake and action. Williams and Wilkins, Baltimore
2. Dudziak R (1980) Lehrbuch der Anästhesiologie. Schattauer, Stuttgart
3. Dick W (1985) Kombinationsanästhesie. Springer, Berlin Heidelberg New York Tokyo (Klinische Anästhesiologie und Intensivtherapie, vol 29)
4. Frankenberger H (1985) Monitoring während Narkosen mit reduziertem Frischgasfluß. In: Lawin P et al (eds) Alternative Methoden der Anästhesie. Thieme, Stuttgart (INA, vol 50)
5. Spiess W (1983) Low Flow and Minimal Flow: Methoden für die Routine. Anaesthesist 32 [Suppl]: 104

6. DIN 13252 (1983) Inhalationsnarkosegeräte - Sicherheitstechnische Anforderungen und Prüfungen. Beuth, Berlin
7. Medizingeräteverordnung (1985) Verordnung über die Sicherheit medizinisch-technischer Geräte (Medizingeräteverordnung). Bundesgesetzblatt p 93
8. Schreiber P (1972) Anaesthesia equipment. Springer, Berlin Heidelberg New York (Anaesthesiology and resuscitation, vol 59)
9. Wawersik J (1982) Die Geschichte des Narkoseapparates in Grundzügen. Anaesthesist 31:541
10. Lowe HJ, Ernst EA (1981) The quantitative practice of anesthesia. Williams and Wilkins, Baltimore

Quantitative Anaesthesia in the Low-Flow System

J. Baum

Introduction

Reduction of the fresh gas flow to those volumes taken up by the patient, that is, the practice of quantitative anaesthesia, would require a fundamental rethinking on the part of the anaesthetist.

In the majority of anaesthesias up till now - even where rebreathing systems are used - fresh gas is delivered in such excess that the rebreathed volume is negligible and the unused and thus useless excess gas volume ventilated from the system is correspondingly large. This practice is justified with the argument that the safety of the patient can only be secured in this manner. As a consequence, the patient breathes in a narcotic gas, the composition of which is independent of the current uptake of oxygen, nitrous oxide, and volatile anaesthetic, and which corresponds essentially to that of the fresh gas. The patient floats, so to speak, in a cloud of fresh gas from which unknown volumes of oxygen, nitrous oxide, and volatile anaesthetic are taken up into the blood, their magnitude depending on the momentary alveolar-arterial partial pressure gradient.

This method of anaesthesia seems even more questionable since the majority of available anaesthesia apparatus are equipped with a technically sophisticated rebreathing system, the circle system, and according to the specifications for medical equipment in Germany (Medizingeräteverordnung-MedGV) they have to be outfitted with technical safety features which are specifically designed for the rebreathing mode [3, 5].

The advantages of the reduction of fresh gas volumes and the consequent use of rebreathing are:

- An improvement in the warming and humidifying of the anaesthetic gases, which results in an improvement in the tracheobronchial atmosphere and a reduction in cooling of the patient
- A reduction of excess gas volumes and a consequent reduction in environmental pollution by halogenated hydrocarbons and nitrous oxide
- A not insignificant reduction in costs due to the more efficient use of the anaesthetic gases

Technical Requirements of the Equipment Necessary for Quantitative Anaesthesia

The fresh gas can only be reduced to the gas volume that the patient takes up if the apparatus fulfills the following technical criteria: the breathing system must be completely gas-tight, the gas-dosaging system must provide adequate precision, the dosaging system for the inhalation anaesthetics must be able to provide sufficient concentrations at low flow rates, and the construction and function of the ventilator must be adapted for use with low fresh gas flow rates. If these requirements are met and the volume and composition of the gases circulating in the system held constant in a steady state, one should be able to determine the amounts of oxygen, nitrous oxide, and volatile anaesthetic taken up by the patient directly from the volume and composition of the fresh gases fed into the breathing system. However, the levels of precision and performance necessary for the practice of quantitative anaesthesia are not met by the anaesthesia machines presently available [1].

Furthermore, the majority of the industrially produced anaesthesia apparatus available have been conceived and licensed for use in the half-closed mode, but not for closed-circuit operation. Since the user is legally obliged to use the apparatus only in the mode for which it was designed and licensed by the manufacturer, not only technical insufficiencies, but also medicolegal considerations speak against the use of these systems for quantitative anaesthesia.

Thus, only those modes of anaesthesia using reduced fresh gas volumes can be realized which, despite the maximal use of rebreathing, employ half-closed operation and at least a minimum of excess gas.

Anaesthesia Techniques Using Low Fresh Gas Volumes in Half-Closed Rebreathing Systems

The two methods of anaesthesia using the half-closed system under maximal exploitation of the rebreathing technique are the low-flow method (described by Foldes in 1952) and the minimal-flow method (introduced by Virtue in 1974). The fresh gas volume is reduced in these methods to 1 l/min and 0.5 l/min respectively [10, 11, 19]. Implementation of minimal-flow anaesthesia entails the maximum possible reduction in fresh gas volume and exploits the industrially produced apparatus to their technical limits. Thus, this method representing an extreme type of half-closed use of a rebreathing system will be the focus of the remainder of the present report.

Minimal-flow anaesthesia according to Virtue is used routinely in our department and is carried out as follows. Premedication with atropine and midazolam is followed by intravenous application of a hypnotic to induce anaesthesia. After preoxygenation with a high oxygen flow, a muscle relaxant is given and the patient is intubated and connected to a rebreathing circle system. In the first 15–20 min and beginning as soon as possible, mechanical ventilation is begun with a high fresh gas flow into the system of 4.4 l/min (1.4 l/min O_2, 3.0 l/min N_2O). During this initial phase of the anaesthesia, nitrogen is washed out of the system, the gas concentrations in all parts of the system are homogenized, the level of anaesthesia is quickly deepened, and the initially existing volume imbalances are bridged.

After this initial phase, which is characterized by the setting of high fresh gas volumes, the oxygen flow is reduced to 300 ml/min, the nitrous oxide flow to 200 ml/min, which is standard, and the fresh gas concentration of the volatile anaesthetic is simultaneously increased.

Depending on the duration of the anaesthesia, the addition of the volatile anaesthetic to the continuing low fresh gas flow may be ceased 15-30 min before the end of the operative procedures. Thereafter, respiration is manually assisted until spontaneous respiration takes over. Immediately prior to extubation, the anaesthesia system is washed out with a high oxygen flow. Monitoring of the patient in the early postoperative recovery phase is carried out according to usual procedures [4].

Minimal-Flow Anaesthesia – A Quantitative Method?

The concept of minimal-flow anaesthesia is based on a standardization of the fresh gas volume setting at 0.5 l/min and the fresh gas composition at 60 vol% O_2 and 40 vol% N_2O. While intraoperative oxygen consumption may be viewed as an approximately constant value varying about 10% ([20, 21]; see also Fig. 1A), the uptake of nitrous oxide falls as an exponential function of time over the course of the anaesthesia ([16]; see also Fig. 1B). The total volume of gas taken up by the patient falls in parallel with the fall in nitrous oxide uptake (Fig. 1C). However, since the fresh gas inflow remains constant over the entire duration of the anaesthesia, the proportion of excess gas vented from the system

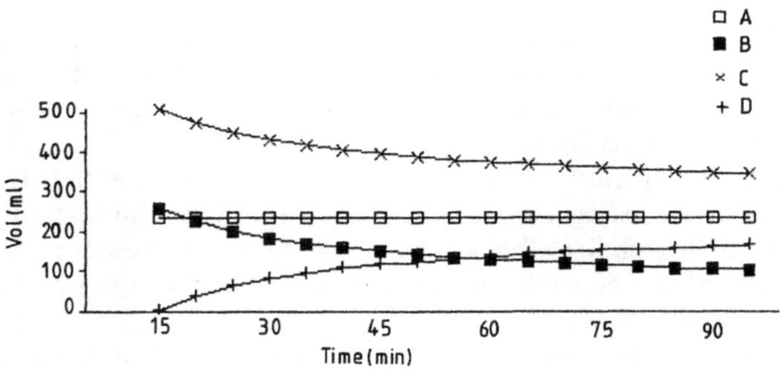

Fig. 1. Uptake and excess gas volume; mathematical simulation of a minimal-flow anaesthesia. Patient's body weight 75 kg, isoflurane anaesthesia, nitrous oxide concentration 65 vol%, fresh gas flow 500 ml/min.
A: Oxygen consumption, calculated according to the formula given by Brody [6]. *B:* Nitrous oxide uptake, calculated according to the formula given by Severinghaus [16, 18]. *C:* Momentary total gas volume taken up by the patient; uptake of isoflurane calculated according to the formula given by Lowe [14]. *D:* Calculated excess gas volume being vented from the system at any given moment

increases continually (Fig. 1D). The continual changes in gas concentrations in the anaesthesia system observed over the course of the anaesthesia are due to the standardization of the fresh gas composition.

This and the continual increase in the excess gas volume due to the constant inflow of fresh gas make it impossible to precisely determine the gas volumes taken up by the patient, on the basis of the apparatus settings. Thus, the concept of minimal-flow anaesthesia itself is incompatible with the principle of quantitative anaesthesia.

Moreover, the technical inadequacies of the dosaging system and the inaccuracy of the measurement instruments are also obstacles to an exact determination of the gas volumes taken up by the patient.

Nonetheless, observation of the gas composition and the volume of circulating gas gives more information about the patient when minimal-flow anaesthesia is used, than when anaesthesia is carried out with high fresh gas volumes.

Oxygen Concentration in the Anaesthetic Breathing System

In minimal-flow anaesthesia, the inspiratory oxygen concentration is much more strongly influenced by the patient's oxygen consumption than in anaesthesia with high fresh gas flows [20]. If a patient's oxygen consumption is high, the oxygen content of the expiratory gas is consequently low, and, as the high proportion of rebreathing mixes large volumes of the oxygen-poor expiratory gas with the only small volumes of oxygen enriched fresh gas, the inspiratory oxygen concentration falls noticeably (Fig. 2B). In contrast, with high fresh gas and low rebreathing volumes, the lower the proportion of rebreathed gas in the inspiratory mixture, the more closely the inspiratory oxygen concentration approaches that of the fresh gas, independent of the patient's oxygen consumption (Fig. 2A).

Our own investigations have also been able to verify that in minimal-flow anaesthesia the inspiratory oxygen concentration is influenced to a clinically relevant degree by the patient's characteristics which correlate with the individual

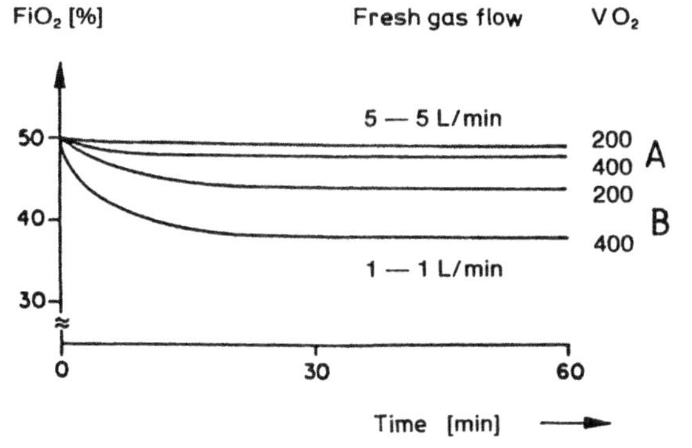

Fig. 2. At a low gas flow (*B:* 1 l/min O_2 and 1 l/min N_2O), increased oxygen consumption (VO_2) results in a significantly greater reduction in the inspiratory oxygen concentration than at a high flow (*A:* 5 l/min O_2 and 5 l/min N_2O) [20]

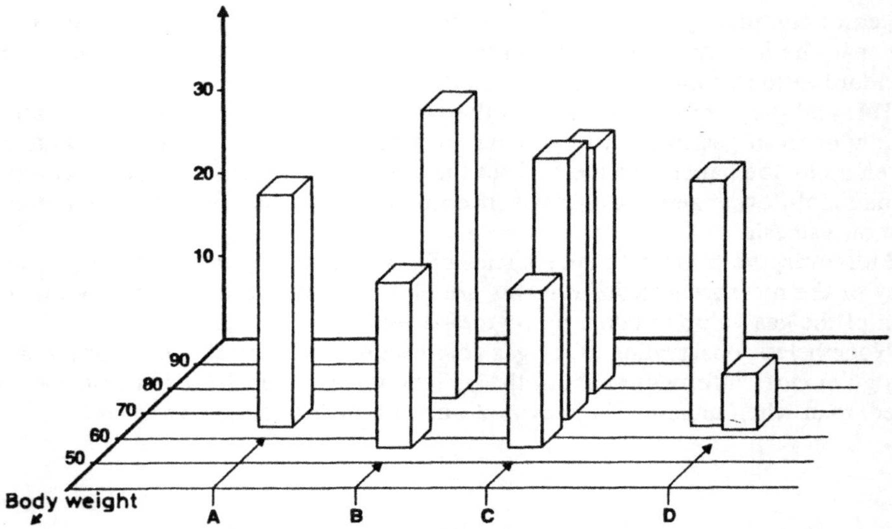

Fig. 3. Relative frequency with which the oxygen concentration fell to the critical value of 28 vol% according to the patient's characteristics: body weight, constitution, and age. *Ordinate*, relative frequency (%); *depth axis*, body weight (kg).
A: All subjects together. Average body weight, 69 kg; average age, 41.5 years. Differentiated according to *B:* average body weight, 59 vs 78 kg; *C:* constitution according to the Broca index (underweight, average 58 kg; normal, average 67 kg; overweight, average 79 kg); *D:* age (average age 38 years vs average age 62 years, for almost identical body weight and stature)

oxygen consumption (Fig. 3). Minimal-flow anaesthesia was carried out with a fresh gas composition of 50 vol% O_2 and 50 vol% N_2O. The inspiratory oxygen concentration increased during this procedure in a typical manner during the initial 30–45 min after flow reduction and sank thereafter slowly but continuously. Whenever the inspiratory oxygen concentration fell below 28 vol%, the composition of the fresh gas was altered to contain more oxygen. In Fig. 3, the bars represent the percentage of anaesthesia in which the inspiratory oxygen concentration fell to this threshold value within a time span of 60 min after reduction of fresh gas flow. This was more frequently the case for heavier patients (B), athletic and obese (C), and young patients (D), than for patients of low body weight (B), leptosomic stature (C), or advanced age (D). Thus, although the exact oxygen consumption can not be determined from measurements of inspiratory oxygen concentration and its changes in the course of minimal-flow anaesthesia, a coarse qualitative estimation is at least possible.

In the past year, oxygen measuring devices have become available which function according to the paramagnetic principle. With these instruments, which are characterized by short response times, it is possible to measure the inspiratory as well as the expiratory oxygen concentration from breath to breath. However, even using these devices, the oxygen consumption can only be determined from the difference between inspiratory and expiratory oxygen volumes, not alone from the concentration differences:

$$\mathrm{UPTAKE}\ (O_2) = \mathrm{VOL(in)} \cdot \mathrm{CON}(O_2\mathrm{in}) - \mathrm{VOL(ex)} \cdot \mathrm{CON}(O_2\mathrm{ex})$$

where

VOL(in) is the inspiratory gas volume;
CON(O_2in) is the inspiratory oxygen concentration;
VOL(ex) is the expiratory gas volume; and
CON(O_2ex) is the expiratory oxygen concentration.

Therefore, in order to determine oxygen uptake in this manner, one would have to measure the inspiratory and expiratory volumes, too. Given the small volume differences involved and the inaccuracy of presently available volumeters (8%–10% of the measured value), it would seem that adequate measures are beyond the scope of clinical practice. Further, the expiratory gases would have to be homogenized in a mixing chamber before the concentration is measured. Finally, this technique for determination of oxygen consumption is not specific to minimal-flow anaesthesia. If it were possible, it would be equally applicable to high-flow anaesthesia. Thus, measurement of the difference between inspiratory and expiratory oxygen concentration alone does not represent a simple method for measuring oxygen consumption.

Since the oxygen consumption can not be exactly determined during minimal-flow anaesthesia, it is also not possible to determine cardiac output in clinical routine simply by applying Fick's principle to oxygen consumption values:

$$\mathrm{CO} = \mathrm{UPTAKE}(O_2)/\mathrm{Ca}(O_2) - \mathrm{Vc}(O_2)$$

where

CO is the cardiac output; UPTAKE(O_2) is the oxygen consumption; Ca(O_2) is the arterial oxygen content; and Cv(O_2) is the mixed venous oxygen content.

Expiratory Carbon Dioxide Concentration

In anaesthesia with reduced fresh gas volumes, the determination of expiratory carbon dioxide concentration is alleged to be of special significance for the quantitative registration of physiological changes [8]. It is known that this parameter provides the anaesthetist with a very complex piece of information which is influenced by respiratory minute volume, cardia output, and metabolic rate [12, 17].

But in complete contrast to the measurement of oxygen and volatile anaesthetic concentrations which, due to the long time constant of a rebreathing system used for minimal-flow anaesthesia, change only with long delays, the measurement of CO_2 concentration allows the immediate registration of physiological changes from breath to breath. It must be mentioned, however, that the measurement of expiratory carbon dioxide concentration provides no new method-specific monitoring possibilities over and above those realizable with higher fresh gas flows or even the half-open mode.

On the contrary, according to our own studies, the following reservation must be made. In carrying out minimal-flow anaesthesia with ventilators employing a hanging bellows and continual inflow of fresh gas into the breathing system, method-specific changes do occur in the measured signal (Fig. 4). Given this equipment configuration, the reduction of fresh gas volume from 4.4 to 0.5 l/min results in changes in the shape of the capnographic curves registered both in the main and side stream measurement technique: an expiratory plateau is not reached as clearly as when high flows are used (Fig. 4: A1, B1, C1), but after peaking the curve shows a more or less decreasing shape during the expiratory phase (Fig. 4: A2, B2, C2). The aeDCO$_2$, the difference between the arterial and expiratory CO$_2$-partial pressure, increases by 2.5 mm Hg on average, and the precision with which the measurement reflects alveolar CO$_2$ concentration is reduced (Fig. 5).

This method-specific change in the measured signal occurring with the configuration described above is due to the following factors:

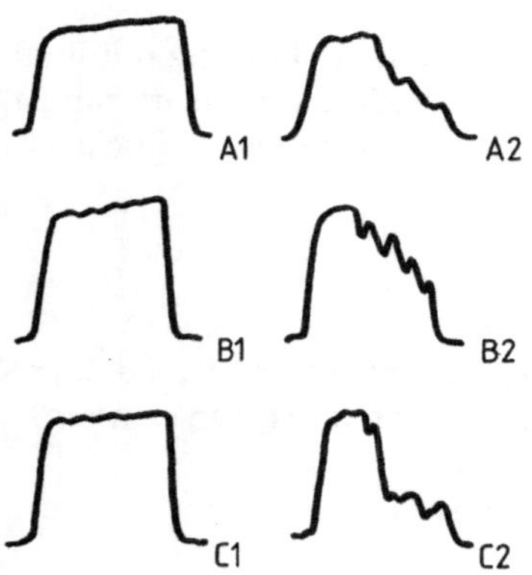

Fig. 4. Method-specific changes in the shape of capnographic curves during ventilation by ventilators with hanging bellows and continuous fresh gas flow into the breathing system.
A1, B1, C1: capnogram with high fresh gas flow;
A2, B2, C2: capnogram with low fresh gas flow

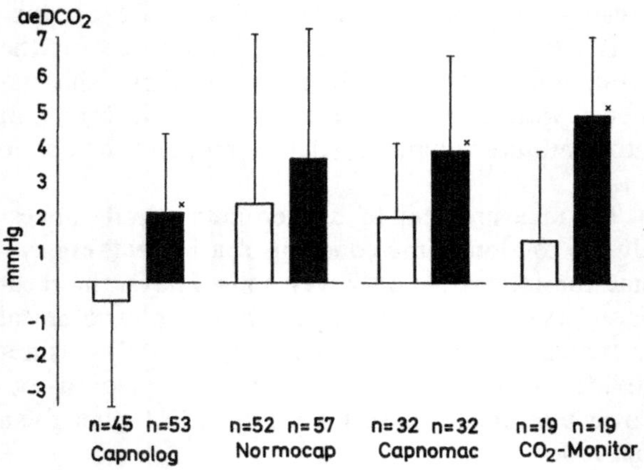

Fig. 5. Increase in the aeDCO$_2$ at the transition from high *(open bars)* to low *(filled bars)* fresh gas flow, measured by main stream (Capnolog, Dräger) and by side stream measurement technique (Normocap and Capnomac, Datex; CO$_2$-Monitor, Dräger); *, difference significant at the 0.001 level

- The flow reduction results in a decrease in the tidal volume with corresponding changes of the gas distribution in the lungs.
- In order to improve the gas tightness of the ventilator, the PEEP valve is switched on so to prevent gas loss via the excess gas valve. The resulting positive end-expiratory pressure, even if low, leads to a corresponding increase in alveolar dead space ventilation.
- The flow reduction furthermore leads to an alteration of the gas flow within the circle system followed by a contamination of the expiratory with the inspiratory gas in the vicity of the tracheal tube's connector.
- When the side stream measurement technique is used, this contamination of the expiratory gas with inspiratory gas can be increased by the suction of the gas-probe removal.

Concentration of the Volatile Anaesthetics

Just as the volume and composition of the fresh gas can be set to standard values in minimal-flow anaesthesia, the concentation of the volatile anaesthetic can also be standardized. This simplifies the anaesthesia method significantly. If the isoflurane concentration in the fresh gas is standardized at 1.5 vol% during the initial phase of anaesthesia in which a higher fresh gas flow is used, and if this concentration is raised to 2.5% when the fresh gas volume is reduced to 0.5 l/min, expiratory concentrations of 0.9–0.8 vol% are obtained (Fig. 6). Together with the MAC fraction represented by a nitrous oxide concentration of 60%–70%, an additive total anaesthetic concentration of 1.3 MAC results which corresponds to the AD_{95} [2]. The same procedure can be used with the inhalation

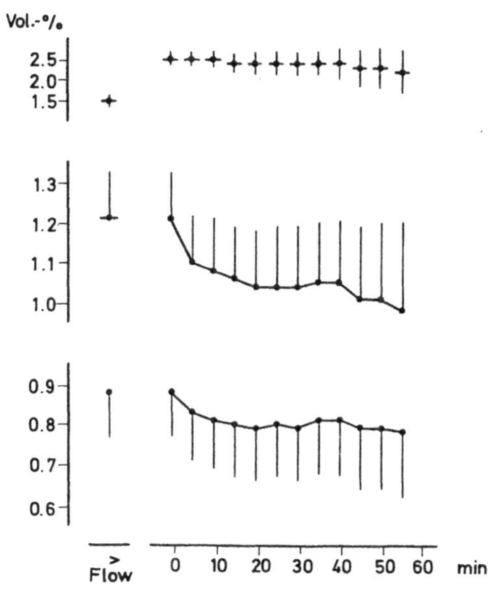

Fig. 6. Inspiratory and expiratory isoflurane concentrations directly before and in the 60 min following flow reduction to 0.5 l/min (mean and standard deviation). *Ordinates: above,* evaporator setting = fresh gas concentration; *middle,* inspiratory isoflurane concentration; *below,* expiratory isoflurane concentration

anaesthetic ethrane, in which case the vaporizer setting may be standardized at 2.5 to 3.5 vol%. Depending on the operative procedures and the individual's reactions, further changes in the fresh gas concentration of the volatile anaesthetic may be made, as when working with high fresh gas volumes. One must keep in mind, however, that the effects of changes in fresh gas concentration are significantly delayed, due to the long time constants of the breathing system in minimal-flow anaesthesia. Even though it is possible to increase the depth of anaesthesia by supplementary doses of analgesics or hypnotics, a rapid decrease of the anaesthetic planum can only be achieved by raising the fresh gas flow and simultaneously reducing the concentration of the volatile anaesthetic.

Our own studies have shown that when standardized fresh gas concentrations are chosen, the concentrations of isoflurane and ethrane established in the anaesthesia system are only insignificantly influenced by such patient characteristics as weight, constitution, and age. The differences are irrelevant, both in their magnitude (0.1–0.2 vol%) and with respect to clinical aspects. Thus, measurement of the concentration of the inhalation anaesthetic and observation of changes in this concentration during anaesthesia do not allow deductions with respect to the individual uptake (Figs. 7 and 8).

Secondly, with low fresh gas flows, the inaccuracies of the gas-dosing systems and the vaporizers are together so large that it is not possible, on the basis of the device settings, to calculate the volumes of volatile anaesthetic fed into the system exactly.

Finally, the concept of minimal-flow anaesthesia assumes fresh gas volumes and concentrations which are standardized and as far as possible constant; that is why the alveolar anaesthetic concentration and the portion of volatile anaesthetic escaping the system with the excess gas continually changes.

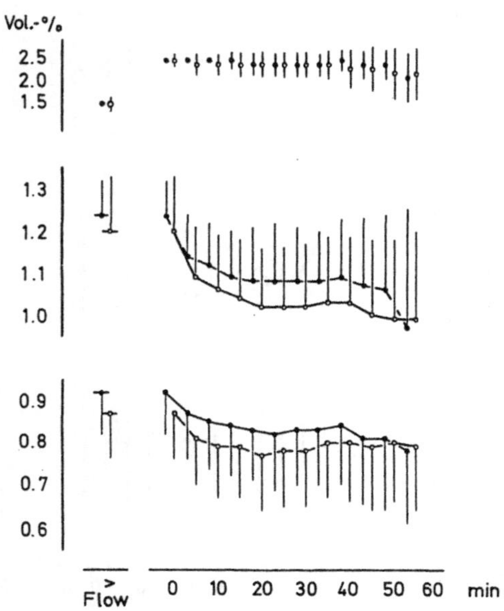

Fig. 7. Isoflurane concentrations in the course of minimal-flow anaesthesia (ordinates as in Fig. 6).
Differentiation according to body weight: *filled circles*, average body weight of 62 kg; *open circles*, average body weight of 81 kg

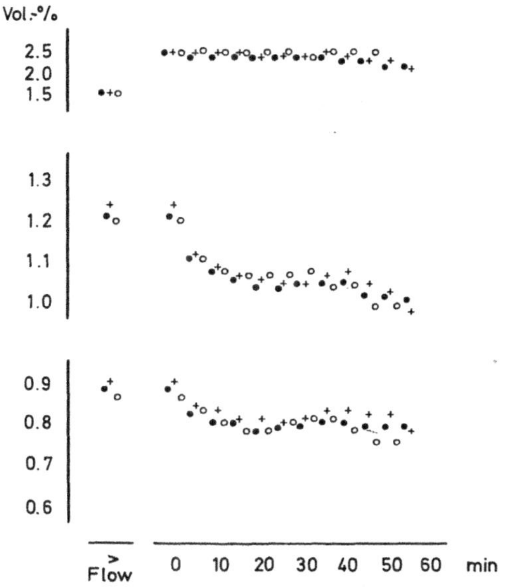

Fig. 8. Isoflurane concentrations in the course of minimal-flow anaesthesia (ordinates as in Fig. 6).
Differentiation according to body stature as quantified by the Broca index: *filled circles*, adipose; *open circles*, underweight; *crosses*, normal

The amounts of inhalation anaesthetic taken up by the patient in minimal-flow anaesthesia could only be determined exactly if: (a) the dosing systems operated to exactly in the low-flow range that the vapor volumes fed into the breathing system could be calculated; (b) the alveolar anaesthetic concentration reached a stable value in a steady state; and (c) the volumes of anaesthetic escaping with the excess gas were known. Since these conditions are not achieved in clinical practice, the dosaging of inhalation anaesthetics in minimal-flow anaesthesia is not carried out in a quantitative manner (i.e., according to the patient's uptake) but according to clinical criteria.

Summary

Minimal-flow anaesthesia represents an extreme mode of operation of the half-closed technique, the technique for which the vast majority of industrially produced anaesthesia apparatus are conceived. Anaesthesia with a half-closed rebreathing system is characterized by the fact that fresh gas is added in excess, that is, in larger amounts than the patient takes up. In order to insure constancy of the volume and pressure within breathing system, excess gas, the volume and composition of which are unknown to the anaesthetist, must be vented continually. If the concept of quantitative anaesthesia is understood to mean that only those volumes of fresh gas are added to the system which are definitely taken up by the patient at any given moment, then neither low-flow nor minimal-flow anaesthesia are quantitative techniques, but rather excess methods.

On the other hand, the extreme reduction of fresh gas volume in minimal-flow anaesthesia (to 0.5 l/min) diminishes the excess volume to the point that the composition of the gases in the breathing system are to a great degree determined by the amounts of the gases taken up by the patient. Thus, this mode of anaesthesia does exhibit the characteristics of quantitative anaesthesia to an extent, insofar as, when the composition and volume of the fresh gas are standardized, the individual oxygen consumption influences the inspiratory oxygen concentration to a clinically relevant degree.

After reducing the fresh gas volume to 0.5 l/min, an initial increase in the inspiratory oxygen concentration is observed which lasts approximately 30–45 min and which is then followed by a continuous decrease (Fig. 9). These changes in inspiratory O_2 concentration result from (a) the equilibration of the system volume with the increased fresh gas oxygen concentration and (b) the exponential fall in nitrous oxide uptake in the temporal course of the anaesthesia, during which the oxygen consumption remains almost constant, and the resulting accumulation of N_2O in the breathing system.

Despite a constant concentration in the fresh gas over the course of anaesthesia, the inspiratory and expiratory concentrations of volatile anaesthetic first fall, and after about 30 min begin to rise very slowly (Fig. 10). This is at least partially due to an exponential decrease in the uptake of the inhalation anaesthetic: with constant inflow, the amount fed into the breathing system is not enough to cover the uptake corresponding to the alveolar concentration established during the initial high-flow phase, and the inspiratory concentration falls. If the uptake then decreases further during the following course of anaesthesia while standardized amounts of anaesthetic agents are continued to fed into the system, the inhalation anaesthetic accumulates and the concentration must rise.

In summary it can be ascertained that changes in the gas concentrations within the breathing system, which are to be observed during minimal-flow anaesthesia, are correlated to the uptake of oxygen, nitrous oxide, and volatile anaesthetics.

In the temporal course of the anaesthesia and due chiefly to the decrease in uptake of nitrous oxide, the volume of the excess gas released from the system

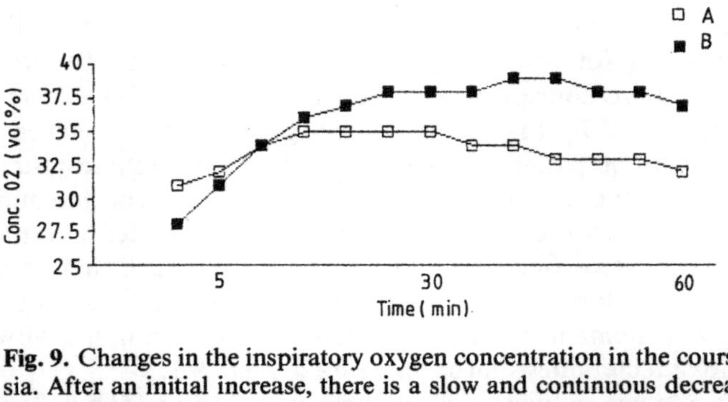

Fig. 9. Changes in the inspiratory oxygen concentration in the course of minimal-flow anaesthesia. After an initial increase, there is a slow and continuous decrease in the O_2 concentration. *Filled symbols,* older patient with a low oxygen consumption; *open symbols,* stronger, younger patient with a correspondingly high oxygen consumption

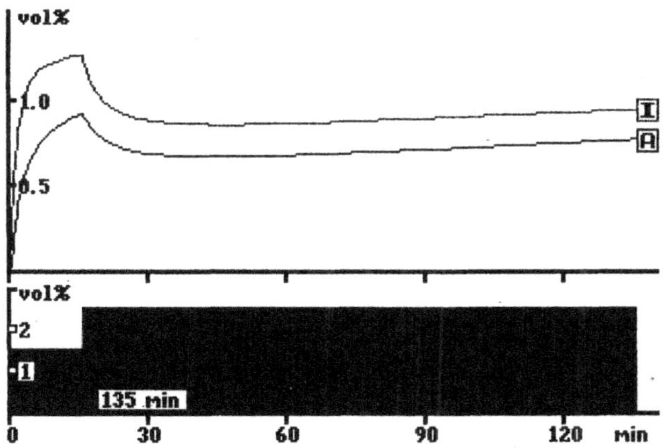

Fig. 10. Computer simulation of a minimal-flow isoflurane anaesthesia lasting 135 min on a patient weighing 75 kg. After reduction of the fresh gas volume there is an initial significant fall in the inspiratory *(I)* and alveolar *(A)* isoflurane concentrations. This is followed by a slow and continual increase in both variables over the further course of the anaesthesia. (Program, [15])

increases so that the characteristics of a half-closed system come to the fore, whereby a dynamic equilibrium in the gas concentration is established. However, all changes in gas composition occur so slowly that a technique-specific danger of hypoxia or overdosaging of the inhalation anaesthetic can be avoided by adequate monitoring of the patient.

Minimal-flow anaesthesia represents a practical compromise between the use of apparatus in adequate for quantitative anaesthesia and the desire to reduce fresh gas volumes to a minimum. This procedure fulfills the requirements for simple operation and practicality, as the volume and composition of the fresh gas are set to standardized values and must not be continually adjusted on the basis of momentary uptake [13].

Thus, the procedure which is recommended here leads away from quantitative anaesthesia:

- In the initial phase, during which large volumes of nitrous oxide and anaesthetic vapors are being taken up and the gas concentrations in the system are approaching an equilibrium, rebreathing is kept small and large excess gas volumes are used.
- The composition and volume of the fresh gas can be established according to a fixed scheme and at such values that changes in the gas composition in the breathing system due to the decrease in gas uptake and the increase in the excess gas volume are kept at tolerable levels over a long time period. But this advantage of minimal-flow anaesthesia is accompanied by a continuous variation of the alveolar gas concentration during the course of anaesthesia, only approximating the desired values and, during particular phases of the anaesthesia, diverging substantially from them.
- The setting of a fixed fresh gas volume results in an increase in the excess gas volume released during the course of the anaesthesia.

Thus, the basic concept of minimal-flow anaesthesia is much more appropriate for fulfilling the need for practicality and ease of operation than is the ambitious approach of quantitative anaesthesia.

One critical comment must be made with respect to quantitative anaesthesia, which conceptually assumes that gas concentrations are the quantities to be regulated in anaesthesia. This is certainly justified in the case of oxygen and nitrous oxide concentrations, as the desired concentrations are in the interest of the patient's safety; this is not, however, so for the dosaging of the inhalation anaesthetics, where the proper criterion for inhalation anaesthesia is that depth which is adequate for the individual patient and operative intervention [7]. The establishment of a defined alveolar anaesthetic concentration can only be seen as a possible starting point which requires correction on the basis of the individual's reaction and the characteristics of the operation in question. The criterion for dosaging of a volatile anaesthetic, even when quantitative anaesthesia is carried out, can not be a particular gas concentration but must be a depth of anaesthesia appropriate to the particular situation, and this may only be judged on the basis of clinical parameters.

Thus, anaesthesia procedures using low-flow systems lead in the direction of quantitative anaesthesia but, out of consideration for better practicality, stop short of this goal. Furthermore, they are the techniques by which a compromise between the desire for a maximal reduction of fresh gas volume and the necessity of using apparatus inadequate for quantitative methods may be achieved. Aside form the advantages listed at the beginning, a significant benefit of these "semi-quantitative" methods which has not been mentioned up to now is that the anaesthetist must familiarize himself with the fundamentals of quantitative anaesthesia [9]. This expanded knowledge of the clinical and technical details of anaesthesia serves the interests of patient safety.

References

1. Baum J (1985) Technische Voraussetzungen für die Narkoseführung mit reduziertem Frischgasfluß. In: Lawin P, van Aken H, Schneider U (eds) Alternative Methoden der Anästhesie, vol 50. Thieme, Stuttgart, pp 43–48
2. Baum J (1986) Minimal-flow anaesthesia with isoflurane. In: Lawin P, van Aken H, Puchstein C (eds) Isoflurane. Anaesthesiologie und Intensivmedizin, vol 182. Springer, Berlin Heidelberg New York, pp 325–331
3. Baum J (1987) Narkosesysteme. Anaesthesist 36:393–399
4. Baum J (1988) Praxis der Minimal-Flow Anästhesie, INA vol 64. Thieme, Stuttgart
5. Bergmann H (1986) Das Narkosegerät in Gegenwart und Zukunft aus der Sicht des Klinikers. Anaesthesist 35:587–594
6. Brody S (1945) Bioenergetics and growth. Reinhold, New York
7. Cullen SC (1972) Who is watching the patient? Anesthesiology 37:361–362
8. Droh R (1986) The closed circuit: past and future. In: Droh R, Spintge R (eds) Closed-circuit system and other innovations. Springer, Berlin Heidelberg New York, pp 3–8
9. Edsall DW (1981) Economy is not a major benefit of close-system anesthesia. Anesthesiology 54:258
10. Foldes FF, Duncalf D (1985) Low flow anesthesia: a plea for simplicity. In: Lawin P, van Aken H, Schneider U (eds) Alternative Methoden der Anästhesie, vol 50. Thieme, Stuttgart, pp 1–7

11. Foldes FF, Ceravolo AJ, Carpenter SL (1952) The administration of nitrous oxide - oxygen anesthesia in closed systems. Ann Surg 136:978-981
12. Lenz G, Klöss T, Schorer R (1985) Grundlagen und Anwendung der Kapnometrie. Anästhesiol Intensivmed 26:133-141
13. Lin CY (1986) Can we practice safe, simple closed-circuit anaesthesia without extensive monitoring or calculus? In: Droh R, Spintge R (eds) Closed-circuit system and other innovations in anaesthesia. Springer, Berlin Heidelberg New York, pp 70-78
14. Lowe HJ, Ernst EA (1981) The quantitative practice of anesthesia. Williams and Wilkins, Baltimore
15. Schwilden H (1986) Abbott-Narkosesimulator: Programm zur Simulation von Konzentrationsverläufen volatiler Anästhetik in verschiedenen Geweben und Organen in Abhängigkeit von der Dosierung, den Parametern des Narkosegerätes und der Beatmung sowie anthropometrischer Daten des Patienten für IBM- und kompatible Computer (Programm- und Datendiskette). Deutsche Abbott, Wiesbaden
16. Severinghaus JW (1954) The rate of uptake of nitrous oxide in man. J Clin Invest 33:1183-1189
17. Smalhout B, Kalenda Z (1975) An atlas of capnography. Kerkebosch, Zeist
18. Spieß W (1985) Sauerstoffverbrauch und Aufnahme von Lachgas und volatilen Anästhetika. In: Lawin P, van Aken H, Schneider U (eds) Alternative Methoden der Anästhesie. Thieme, Stuttgart, pp 8-18
19. Virtue RW (1974) Minimal flow nitrous oxide anesthesia. Anesthesiology 40:196-198
20. Westenskow DR (1986) How much oxygen? Int J Clin Monit Comput 2:187-189
21. Westenskow DR, Jordan WS (1982) Automatic control of closed circuit anesthesia and the measurement of enflurane N_2O and oxygen uptake. In: Abstracts of the International Symposium on Geschlossenes System für Inhalationsnarkosen, Düsseldorf, May 7-8

Quantitative Anaesthesia in Closed Circuit

E. A. Ernst

Introduction

This presentation is included in the session en titled "How can the patient benefit from quantitative anaesthesia practiced with today's equipment?" The argument is presented that closed-circuit anaesthesia practice benefits everyone - the patient, the anaesthesiologist, the manufacturers of monitoring equipment, the developers of anaesthesia systems, and even society in general.

Definition of Closed Circuit

By definition closed-circuit anaesthesia is simply a delivery technique [1]. It is a constant-volume delivery system requiring that the exhaust valve remain closed. The anaesthesiologist must deliver the exact amount of oxygen and anaesthetic gases taken up by the patient during the anaesthetic course. If delivery exceeds uptake, the volume of the reservoir bag increases, and if delivery is less than patient uptake, the resevoir volume will decrease. Rate of delivery, then, becomes a unique value equaling patient uptake plus circuit leaks, if any. What better way to quantitate anaesthesia! Because the anaesthesiologist is required to deliver exactly what the patient consumes, the patient himself becomes the determining factor for the delivery of oxygen and anaesthetic agents.

How Can the Patient Benefit?

High-flow delivery systems offer dehydrated, bone-dry, cold, fresh gases to the trachea breath after breath and the hydrated and warmed exhaled gases are exhausted from the system. During closed-circuit anaesthesia, the warmed and hydrated exhaled gases are retained in the system.

Most importantly, the anaesthesiologist benefits by acquiring additional monitoring information, and this should represent a great benefit to the patient. Because the anaesthesiologist is required to deliver only the amount of oxygen being consumed by the patient, the rate of oxygen delivery needed to replenish the circuit volume is a noninvasive indicator of oxygen consumption. Such information is discarded during high-flow or even low-flow open anaesthetic circuits. Oxygen uptake is related to cardiac output through the Fick principle and a change in the required oxygen delivery during closed-circuit anaesthesia usually signals a similar directional change in cardiac output. During aerobic metabolism, oxygen consumption is related to carbon dioxide production and the alveolar ventilation needed to excrete the carbon dioxide. By measuring oxygen consumption using a closed-circuit oxygen replenishment technique, quantification of other important physiologic variables becomes available [2].

The desirability of identifying molecular species, and their concentrations, in closed circuits accelerated the development of monitoring equipment and devices. Today gas analyzers facilitate the easy practice of closed-circuit anaesthesia. The anaesthesiologist can now know the inspiratory and expiratory concentration of all important gases including oxygen, carbon dioxide, nitrogen, nitrous oxide, and all volatile agents. Appropriate amounts of anaesthetic agents and oxygen can be delivered empirically without an exhaustive knowledge of the square root of time [3] or other pharmacokinetic models. These monitoring devices certainly help the anaesthesiologist deliver better patient care. Another happy beneficiary is the manufacturer of monitoring equipment.

Although the monitoring equipment may add to the cost of anaesthesia, many savings are realized through the use of closed circuits [4–6]. Depending upon the length of the anaesthetic, up to 90% of agents can be saved during closed-circuit anaesthesia. The annual savings would exceed $100 million in the United States alone. The economic advantage is realized by the hospital; hopefully, savings are passed on to the patient. Because of financial constraints, expensive agents such as isoflurane can only be used in a closed circuit in many poor countries today. For the same reason closed circuit was employed to conserve expensive cyclopropane during its era of popularity three decades ago.

The fact that no anaesthetic gases are exhausted into the operating room may benefit operating room personnel. The argument of whether trace anaesthetic gases represent a health hazard has raged over the past two decades. It is still unresolved, but everyone agrees that the inhalation of trace anaesthetic gases is probably not beneficial. Scavenging fluorocarbon anaesthetic agents during high-flow anaesthesia and exhausting them into the atmosphere may contribute to a depletion of ozone in the earth's stratosphere. According to the Molina hypothesis a decrease in the stratospheric ozone allows the earth to be bombarded with damaging ultraviolet ray. Industry contributes most of the exhausted fluorocarbons, but anaesthesiologists practicing high-flow techniques contribute some, and during closed-circuit anaesthesia practice the loss of fluorocarbons to the atomosphere is extremely small. The answer to pollution is not scavenging; it is simply - do not pollute.

Are there any disadvantages? All exhaled gases except carbon dioxide will accumulate in a closed circuit. They include methane, hydrogen, nitrogen, and carbon monoxide [7]. None of these has been shown to be dangerous. Nitrogen could only be harmful if its accumulation threatened the circuit concentration of oxygen. Most practitioners use an oxygen analyzer today. Carbon monoxide is a normal breakdown product of hemoglobin. It will accumulate faster if the patient is a heavy smoker or if a large amount of stored blood is transfused [8]. The prudent anaesthesiologist exhausts the reservoir bag every 2 or 3 h to freshen the circuit and avoid accumulation of unwanted gases [9].

How Difficult Is Closed-Circuit Practice?

Despite clinical uptake models developed by Lowe [2], Westenskow [10], and others, closed-circuit anaesthesia practice has not enjoyed widespread popularity. Those who reject the delivery system, or never try it, say that it is too difficult, and even dangerous. Clinical practice using the square root of time model and Unit Doses has been described in many publications over the past two decades [1, 9, 11, 12]. The models invite a clearer understanding of pharmacokinetics, but the advent of circuit gas analysis allows an empiric practice as easy and safe as open-circuit delivery.

Induction of anaesthesia is achieved, as usual, with high-flow techniques. The circuit is closed by decreasing the delivery flow rate of oxygen to the anticipated metabolic requirement, and of nitrous oxide until the circuit reservoir maintains a constant volume at end-tidal conditions when all exhaust valves are closed. Introdution of volatile agent into the breathing circuit can be achieved with the use of an out-of-circuit tek-type vaporizer, copper kettle, or liquid injection. A tek-type vaporizer is usually opened to its maximal setting, and the concentrations of inhaled and exhaled anaesthetic agent are followed using a gas analysis monitor. If only oxygen is used as a carrier gas, the amount of vapor delivered through a tek-type vaporizer will be insufficient during the first half hour of the anaesthetic and supplemental amounts will have to be administered through a high-efficiency copper kettle vaporizer or by liquid injection. Regardless of the method of volatile agent delivery, the ability to monitor the end-tidal concentration makes the administration of the agent very simple.

A charcoal filter shunt on the inspiratory limb of the circuit allows a very rapid decrease of inhaled concentration [13]. Otherwise, high flows must be employed, and the circuit opened in order to awaken the patient. Emergence following comparable depths of anaesthesia after closed or open delivery is exactly the same.

As with any technique, closed-circuit anaesthesia becomes easier, safer, and more useful with practice. The new monitors shorten the learning period and afford an opportunity to enjoy all the benefits of closed-circuit anaesthesia practice.

Recent Studies Addressing Controversial Issues

Over the past decade there have been many rhetorical and emotional pronouncements about how wonderful and desirable closed-circuit anaesthesia is by those who practice it. The open-circuit practitioners, on the other hand, have condemned the techniques as dangerous and unnecessary. Only recently have data been available examining the questions of safety and efficacy. At the annual meeting of the American Society of Anaesthesiologists in October, 1987 the results of a 3-year prospective patient safety study comparing open and closed circuits were reported [14]. Three hundred anaesthetics were randomly assigned to be delivered by open or closed techniques. Table 1 shows that no significant differences in process or outcome measurements were found. The study con-

Table 1. Comparison of open and closed systems

	Open mean (SE)	Closed mean (SE)	P value for difference
Emergence time (min)	11.25 (1.43)	8.24 (1.46)	0.07
Arrival PARS	8.31 (0.11)	8.57 (0.12)	0.05
RR time (min)	114.15 (6.87)	120.46 (7.06)	0.44
Δ temp. (° C)	– 1.74 (0.11)	– 1.73 (0.10)	0.95
Complications	n = 3	n = 5	0.24
Patient rating	9.58 (0.10)	9.77 (0.10)	0.09
Undesirable CV record (%)[a]	14.85 (3.32)	14.80 (4.44)	0.98

[a] Judged by blinded practitioners

cluded that closed-circuit and open-circuit anaesthesia delivery were equally safe when practiced by anesthesiologists trained in both techniques.

Addressing the problem of inconvenient operation of the closed-circuit system, Ritchie, et al. have just decsribed a closed loop control of anaesthesia delivery system [15]. The servo system closes three loops. First the delivery of oxygen, nitrous oxide, and air are automatically controlled depending upon the volume of the circuit and the patient's exhaled oxygen concentration. Tidal volume and respiratory rate are automatically adjusted according to differences between the anaesthesiologist's desired exhaled carbon dioxide concentration and the measured value. The delivery of volatile anaesthetic agent is then automatically controlled according to differences between the anaesthesiologist's desired end tidal anaesthetic agent concentration and the measured concentration. The equipment was exhaustively tested in animals and is now being used for clinical trials. It provides convenience, accuracy, and documentation. Dr. Westenskow et al. described the automatic delivery of enflurane, oxygen, and nitrous oxide in a closed-circuit system using a patient servo mechanism several years ago [16].

For some time, closed-circuit practioners have argued that, in a leak-free system, oxygen delivery equaled oxygen consumption. The hypothesis had not been tested until recently. Employing the closed-loop system developed by Ritchie, oxygen consumption in pigs, calculated by the Fick principle, was compared with replenishment as determined by the closed-circuit system during different physiologic states [17]. Oxygen replenishment and Fick calculations were compared during controlled conditions, at aortic cross clamping and aortic unclamping. Figure 1 shows a typical response of the delivery to aortic cross clamping. The decrease in oxygen delivery is obvious and immediate. Figure 2 shows the direct correlation between calculated oxyen uptake and measured oxygen delivery. The non-zero y-axis intercept is due to a measured 50 ml/min circuit leak. It should be pointed out that inaccuracies in the Fick calculations contribute to the lack of an even better correlation. The authors conclude that the closed-circuit delivery technique provides a sensitive, noninvasive, and continuous monitoring of oxygen uptake during surgery. It could be added that it also indicates directional changes in cardiac output.

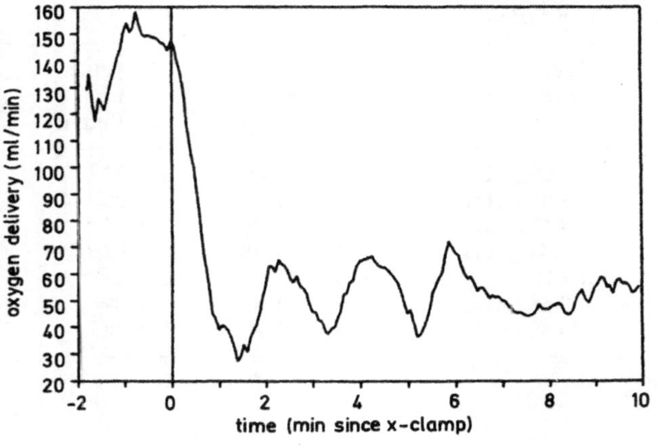

Fig. 1. Typical response of the oxygen delivery controller to aortic cross clamping (at time = 0)

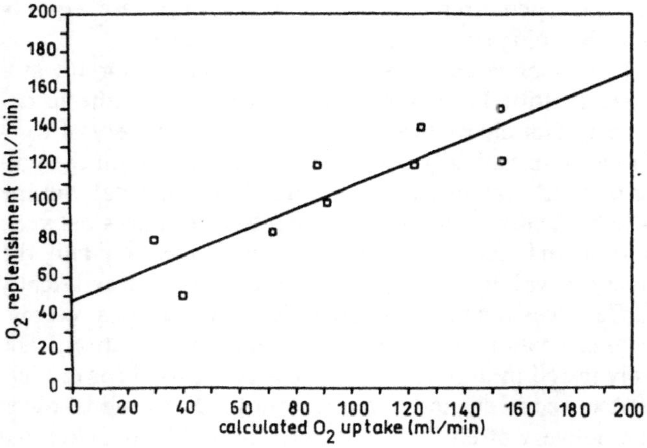

Fig. 2. Correlation (r = 0.88) between automatically controlled oxygen delivery into a closed breathing circuit and oxygen uptake calculated from the Fick equation

Summary

Recent studies seem to vindicate the opinions of those who have been praticing closed-circuit anaesthesia. In a relatively leakless system, oxygen delivery is related to oxygen consumption and all the important physiologic and metabolic variables associated with oxygen consumption. It has been shown that closed-circuit anaesthesia practice does not place the patient at any more risk than open-circuit anaesthesia.

The development of new technology, especially gas analyzers, facilitates the practice of closed-circuit anaesthesia and obviates the necessity of utilizing complicated pharmacokinetic models. Emerging technologies incorporating servo systems will permit each patient to become his own uptake model. Anaesthesiologists will soon control equipment capable of automatically responding to physiologic and pathologic variations within each patient.

References

1. Ernst EA (1985) A clinical approach to closed circuit anaesthesia. Circular 2:5–7
2. Lowe HJ, Ernst EA (1981) The quantitative practice of anesthesia: use of closed circuit. Williams and Wilkins, Baltimore
3. Lowe HJ (1972) Dose related anesthesia (monograph). Abbott Laboratories, Chicago
4. Matjasko J (1987) Economic impact of low-flow anesthesia. Anesthesiology 67:863–864
5. Herscher E, Yaekel AE (1977) Nitrous oxide-oxygen based anesthesia: the waste and its costs. Anesth Rev 4:29–31
6. Spain JA (1981) Cost of delivery of anesthetic gases re-examined: III. Anesthesiology 55:711–712
7. Morita S, Latta W, Hambro K et al (1985) Accumulation of methane, acetone, and nitrogen in the inspired gas during closed-circuit anesthesia. Anesth Analg 64:343–347
8. Anderson EW, Andelman RJ, Strauch JM et al (1973) Effect of low-level carbon monoxide exposure on onset and duration of angina pectoris. Ann Intern Med 79:46–50
9. Ernst EA (1986) Closed and open anesthetic circuits – the pros and cons. ASA Annual Refresher Course Lectures 234-1-234-6
10. Westenskow DR, Johnson CC, Jordan WS et al (1977) Instrumentation for measuring continuous oxygen consumption of surgical patients. IEEE Trans Biomed Eng 24:331–337
11. Ernst EA, Spain JA (1983) Close circuit versus high flow systems: examining alternatives. In: Brown B (ed) Contemporary anesthesia practice. Davis, New York, pp 11–38
12. Ernst EA, Pearson JD (1987) Principles and practice of closed circuit anesthesia. Adv Anesthesiol 89–122
13. Ernst EA (1982) Use of charcoal to rapidly decrease depth of anesthesia while maintaining a closed circuit. Anesthesiology 57:343
14. Ernst EA, MacKrell TN, Pearson JD et al (1987) Patient safety: a comparison of open and closed circuits. Anesthesiology 67:A474
15. Ritchie RG, Ernst EA, Pate BL et al (1987) Closed-loop control of an anesthesia delivery system: development and animal testing. IEEE Trans Biomed Eng 34:437–443
16. Westenskow DR, Jordan WS, Hayes JK (1983) Uptake of enflurane: a study of the variability between patients. Br J Anaesth 55:595–601
17. Ritchie RG, Gelman S, Ernst EA et al (1987) Continuous monitoring of oxygen uptake by replenishment compared to oxygen uptake calculated by the Fick equation. Presented at Computers in anesthesiology VIII, Pine Island Resort, 15 October, 1987

Clinical Requirements of Quantitative Monitoring: Low Flow and Closed Circuit

T. Oyama, H. Ishihara, K. Isozaki, T. Suzuki, and A. Matsuki

Introduction

The continuous measurement of concentrations of gases (including oxygen, carbon dioxide, and inhalation anaesthetic agents) is essential and integral to the safety of closed-circuit anaesthesia.

An oxygen analyzer is required for maintaining adequate oxygen supply, and an end-tidal carbon dioxide monitor is used to forestall accumulation of carbon dioxide in the circuit and for maintaining appropriate ventilation. Measurements of anaesthetic gas concentration such as halothane can not directly determine the depth of anaesthesia. However, once the concentrations are known and integrated with other parameters of anaesthetic depth, such as respiration, the depth itself can be inferred. Furthermore, the difference between the volumes of inhaled and exhaled anaesthetic gases as well as pressure determine the rates of anaesthetic uptake. The pulse oximeter is recommended to monitor tissue oxygen supply in the peripheral blood. The above-mentioned monitoring equipment, measuring many quantitative parameters including blood pressure, ECG, and body temperature are minimal requirements for administering low-flow anaesthesia safely and easily (Table 1).

Oxygen Analyzer

The oxygen analyzer should be placed on the expiratory limb, the site where the lowest concentration of oxygen is to be anticipated. The oxygen concentration

Table 1. Respiratory data

1. O_2, CO_2, halothane analyzer
   F_IO_2, FeO_2, $FetCO_2$, *Fi* halothane, *Fe* halothane
2. Mass spectrometer
   O_2, CO_2, N_2O, N_2, halothane
3. Hot wire respiratory flow meter
   respiratory rate, $\dot{V}e$, $\dot{V}i$, V_T
4. Pulse oximeter
   SaO_2

should be kept at 25%–30%. There are many methods for determining oxygen concentration, including the oxygen electrode and gas mass spectrometer.

Oxygen Electrode Methods

Oxygen electrode methods are based on the ability of oxygen to accept an electron. Two methods are available to measure oxygen concentration including polarographic methods and paramagnetic method. The polarographic electrode oxygen analyzer involves the polarizing voltage between a cathode and a silver anode in an electrolyte solution; at any particular voltage, the current produced varies in proportion to the oxygen tension. Its advantages include rapid and accurate response under very humid conditions; the fact that it is unaffected by carbon dioxide and nitrous oxide; and that it achieves the minimum effects with positive pressure. The disadvantages are that it is battery-powered, requiring regular changes of membrane (once a year) and electrolytes (four times a year).

The paramagnetic analyzer contains a sample cell, a single D cell battery to power a light, and a scale on which the light shadow is cast by a mirror which turns with the changing oxygen concentration. Its advantages include small in size and highly accurate. Its disadvantages are slow response and fragile. Water vapor should be absorbed before analysis to avoid damaging the detector.

Carbon Dioxide Monitor

The sampling tip is connected to the angle piece and the endotracheal tube. End-tidal carbon dioxide is usually kept between 4%–5%.

With the infrared carbon dioxide analyzer, most of the gases and vapors including carbon dioxide have characteristic absorption bands in the infrared region of the spectrum. The infrared carbon dioxide analyzers used in anaesthesia are equipped with cells containing nitrous oxide in the sample. The advantages here are accuracy and rapid response time. The disadvantages are that the samples have to be drawn from the respiratory system and the analyser needs time to warm.

Halothane Analyzer

Using the ultraviolet halothane meter, halothane absorbs ultraviolet light with a peak absorption in the region of 200 nm. It is reasonably accurate and has rapid response time. However, it requires preheating time and the samples have to be drawn from the system, usually at a normal flow rate of 50 ml/min.

With the physicochemical method, the elasticity of the silicon rubber bands relaxes in the presence of halothane vapor. This equipment is simple, provides reasonable accuracy, is independent of an electrical source, and requires no preheating. However, its response time is slow and it is influenced by the presence of nitrous oxide. It is not specific for halothane.

Gas Mass Spectrometer

The gas mass spectrometer measures the composition of inspired and expired gases: oxygen, carbon dioxide, nitrogen, and anaesthetic gases including halothane and enflurane. In addition, nitrous oxide can be measured on a simultaneous breath-by-breath basis. It is an advanced method for accurately and easily measuring gas concentration during closed-circuit anaesthesia.

It operates on the principle that at the pericapillary area, where the pressure has been reduced, a small proportion of the gas sample is supplied to the analyzer which ionizes it at reduced pressure by an electron beam. These ions are focused into a beam by an electrostatic lens, then passed through a magnetic field. Each ion has a characteristic mass and carries an electric charge. Ion collectors are placed at appropriate locations to collect specific ions. Each ion current is amplified to a level convenient for use with conventional meters or recording devices. An ion vacuum pump maintains the internal part of the mass spectrometer at appropriate pressure and suitable power supplies are provided.

Both the mass spectrometer and a compact flow meter with quick response are required to measure oxygen consumption ($\dot{V}O_2$) and carbon dioxide production ($\dot{V}CO_2$). The hot-wire flow meter is a hot-wire anemometer of a constant-temperature type; it is designed as a respiratory monitor in anaesthesia and intensive care.

Computed data are calculated from the following formulas:

$$\dot{V}O_2 = \dot{V}_I F_I O_2 - \dot{V}_E F_E O_2$$

$$\dot{V}CO_2 = \dot{V}_E F_E CO_2$$

where $\dot{V}_I$ is inspiratory gas flow, $\dot{V}_E$ is expiratory gas flow, F_IO_2 is fractional concentration of inspired oxygen, F_EO_2 is fractional concentration of expired oxygen, F_ECO_2 is fractional concentration of expired carbon dioxide.

We have developed a computerized respiratory monitor system for closed-circuit anaesthesia. The system consists of a mass spectrometer, a hot-wire respiratory flow meter, a microcomputer, two central displays, and a recorder. The respiratory parameters obtained by our computerized system include oxygen consumption, carbon dioxide output, respiratory quotient (*RQ*), expiratory volume (*VE*), and respiratory rate (*RR*; see Fig. 1).In the operating room gas is sampled from the tracheal tube and concentrations of oxygen, carbon dioxide, nitrous oxide, and halothane are displayed on the panel of the mass spectrometer. Output data from a hot-wire flow meter and a mass spectrometer are entered in a 16-bit microcomputer.

Clinical Applications

The effects of low-flow anaesthesia on respiratory metabolism were studied in forty surgical patients who underwent either lower abdominal surgery or knee-

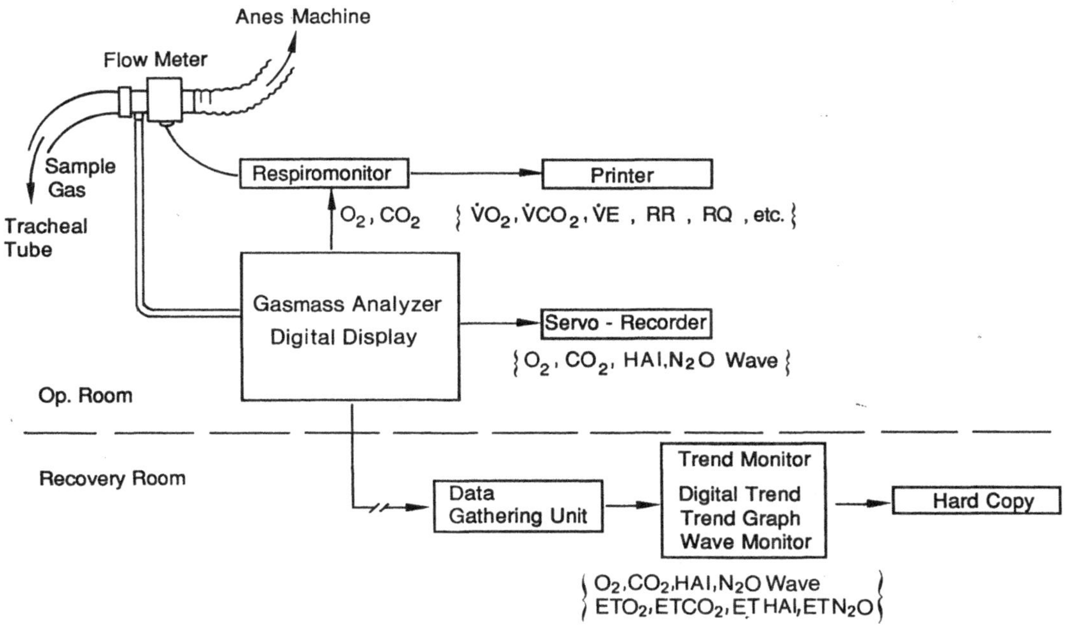

Fig. 1. Diagram of respiratory monitoring system

joint surgery. The patients were randomly divided into two groups; those who were administered low-flow anaesthesia and those who were given conventional semiclosed anaesthesia.

Each patient received diazepam 10 mg i.m. and atropine 0.5 mg i.m. 1 h before anaesthesia was induced with thiopental. Succinylcholine chloride 40 mg was injected to facilitate tracheal intubation. Immediately after the intubation closed-circuit anaesthesia started. The escape valve was shut and fresh gas flow of 250 ml/min oxygen and 500 ml/min nitrous oxygen was set on the flow meter (Fig. 2). Flotec mark 3 was used as halothane vaporizer. The dial of the vaporizer was adjusted according to the end-tidal halothane concentration displayed on a mass spectrometer.

Pancuronium bromide was injected when muscle relaxation was required. Ventilation was assisted manually throughout the procedure and adjusted to keep the end-tidal carbon dioxide concentration around 4.8%. Fresh gas flow of oxygen and nitrous oxide was adjusted to maintain 30% of inspired oxygen concentration during the procedure (Fig. 3).

Oxygen consumption at 30 min after the start of surgery in both surgical groups under low-flow anaesthesia increased by 1.2 to 1.3-fold compared with those at the start of surgery. Carbon dioxide output remained unchanged during the procedure compared with those at the start of surgery in both surgical groups under low-flow anaesthesia. The respiratory quotient decreased slightly following the start of surgery and maintained a steady state ranging between 0.8 and

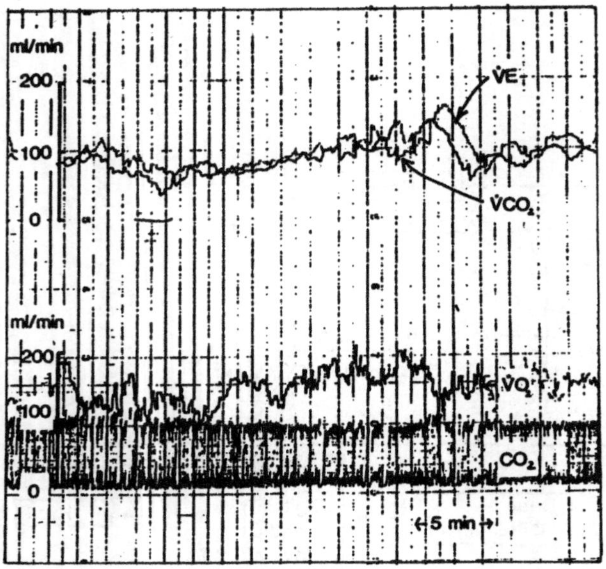

Fig. 2. Measurement of parameters immediately after tracheal intubation of 46-year-old man, body weight 58 kg

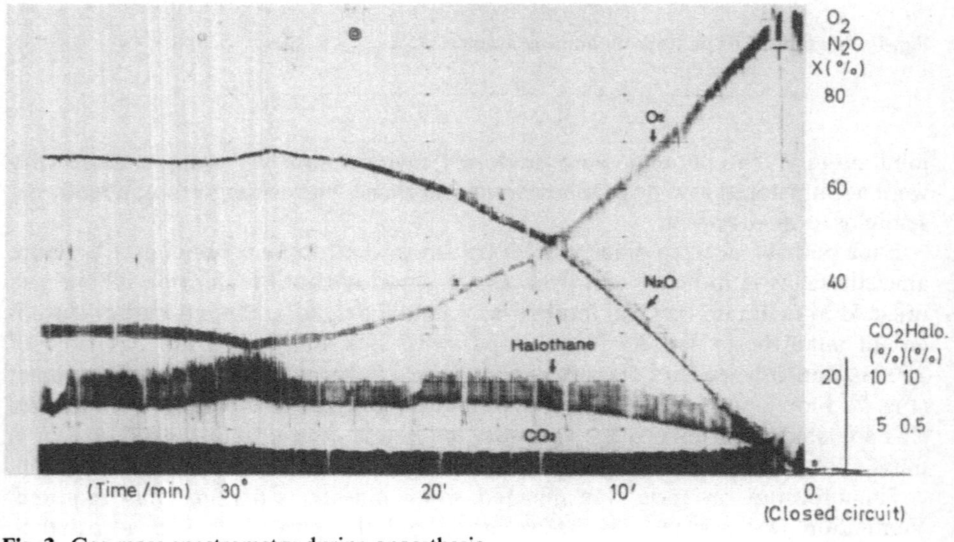

Fig. 3. Gas mass spectrometry during anaesthesia

0.9 in both surgical groups under low-flow anaesthesia (Fig. 4). Oxygen consumption, carbon dioxide output, and respiratory quotient during low-flow anaesthesia were similar to those during semiclosed anaesthesia.

A remarkable difference was observed between the halothane concentration shown by the dial of the Mark 3 halothane vaporizer and that in the anaesthesia circuit, particularly at the start of closed-circuit anaesthesia as shown in this Fig. 5.

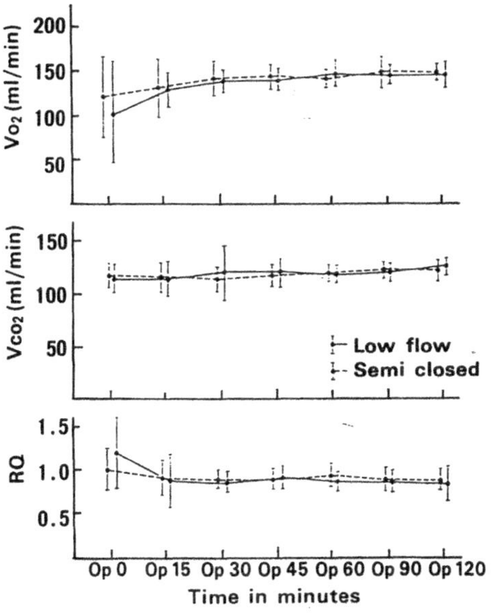

Fig. 4. Measurement of parameters during anaesthesia for lower abdominal surgery (mean ± SE)

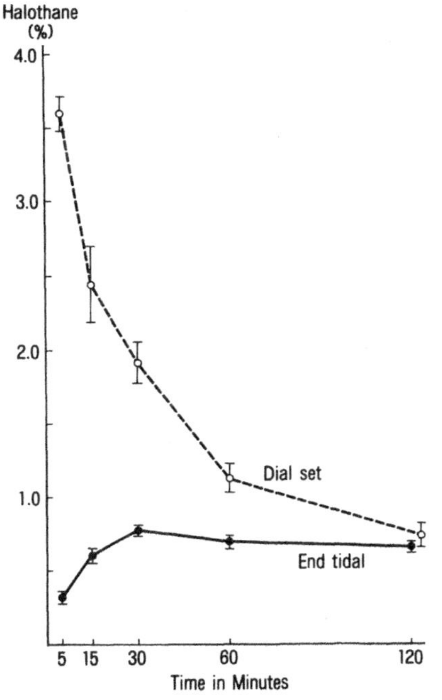

Fig. 5. Halothane concentration during closed-circuit anaesthesia (mean ± SE)

For example, the dial setting of halothane at around 3.5%, at 5 min after the start of anaesthesia showed 0.3% expired halothane concentration. The nitrogen concentration decreased from 5.2% at the start of surgery to 1.2% at 60 min afterwards in the closed-circuit anaesthesia groups. These results suggest that there is no significant difference between the two anaesthetic methods with regard to respiratory metabolism.

The advantage of our method is that it permits the breath-by-breath respiratory metabolism to be evaluated easily, using a simple attachment to the tracheal tube. However, the disadvantage is that the oxygen consumption varies markedly compared with the values for carbon dioxide output. This might be due partly to the frequent changes of inspired oxygen flow or oxygen concentration. Humidity in the circuit also modifies the actual values. The computerized respiratory monitor must be calibrated before each induction of anaesthesia.

Pulse Oximeter

In our opinion the continuous monitoring of blood oxygen saturation is important for the practice of safe and smooth closed-circuit anaesthesia. The spectrophotometry and plethysmography are combined in the pulse oximeter by optically measuring the arterial pulsation with simultaneous red light and infrared light (Fig. 6). The ratio of infrared light absorption (b) to red light absorption (a) exceeding 1 indicates oxygen-saturated blood, while a ratio of less than 1 sug-

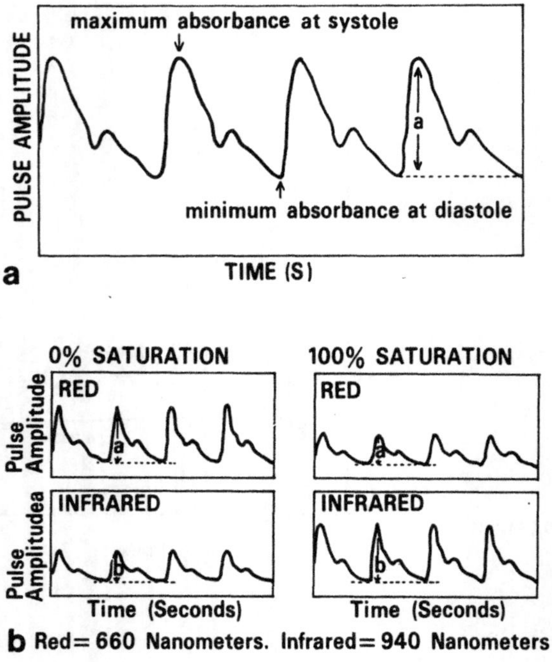

Fig. 6a, b. The pulse oximeter showing the plethysmographic pulse (**a**) and the saturation vs pulse amplitude (spectrophotometry) (**b**)

gests desaturated blood. When the ratio equals 1 the blood is partially saturated (approximately 85% saturation). Arterial blood can be selectively measured independent of capillary or venous blood, and independent of skin color or skin thickness, by observing the pulsatile portion of a light beam through the finger (plethysmography).

The pulse oximeter requires no calibration, no set up. The finger-tip sensor can be conveniently placed away from the head, which is often cluttered with anaesthetic equipment. However, the pulse oximeter can not be used on a patient with intense vasoconstriction, because of the reduced pulsatile flow in the fingers, ears, or other peripheral tissue. Intravascular dyes such as indocyanine green or methylene blue absorb red and infrared light, which causes errors in optical measurement. In addition, an abnormal form of hemoglobin (e.g., methemoglobin) can elicit errors.

Fresh Gas Flow Rate and I:E Ratio Affect Tidal Volume in Anaesthesia Ventilators

J. A. Aldrete, A. J. Adolph, L. M. Hanna, H. A. Farag, and M. Ghaemmaghami

In the past, the general consensus among anaesthesiologists has been that, in most ventilators used during anaesthesia, changes of fresh gas flow (FGF) fed to the ventilator from the anaesthesia machine and into the breathing circuit, do not usually alter the tidal volume (Vt) delivered to the patient, as long as the ventilator settings were unaltered [1, 2]. However, Ghani [3] found that varying the FGF from 2.5 to 10 l/min significantly increased the minute volume. In a preliminary study [4], we were able to confirm his observations indicating that the settings of expected tidal volume (Vt) and rate (f), as well as those of minute volume (V), do not represent the actual volumes exhaled by anaesthetized patients. To determine the validity of the premise, we have investigated several factors that may influence the delivered Vt on several anaesthesia ventilators.

Methods

Changes of Fresh Gas Flow (FGF) Rate

Our initial observations were made by measuring the actual tidal volumes produced by four ventilators commonly used in anaesthesia, when the FGF reduced in a step-down fashion. The ventilators studied were the Narco Air Shields, the 702 Fraser Harlake, the Dräger A-V, and the Ohio 7000. Preliminary observations were also made with a model being developed by Engström-Gambro AB and a modification of the Ohmeda 7800. In all observations the same Fraser Harlake respirometer was attached proximally to the common limb of the Y-piece of the breathing circuit and distally to a corrugated tube, of approximately 100 ml in capacity, attached to a breathing bag to simulate the clinical situation [4].

The actual tidal volumes generated by the four mentioned ventilators were measured, starting with a FGF of 10 l/min which four mentioned ventilators were measured, starting with a FGF of 10 l/min which was then decreased, every 15 min, by 1 liter measurements, down to 1.0 l/min and then to 0.5 and 0.3 l/min, using a constant I:E ratio of 1:2 and a selected Vt of 1 l/min. At each level, the displacement of the bellows was confirmed as being precisely 1000 ml/breath and the frequency 10/min, as it had initially been set. Measurements of peak pressure, the delivered Vt and V were also made, each time.

Variations of I:E Ratio Plus FGF Changes

Similar studies were conducted with two other ventilators, the ECAV Ventilator attached to Foregger 500 anaesthesia machines and the Narkomed 2A Ventilator. Measurements were made not only varying the FGF rate, but also at I:E ratios of 1:1, 1:1.5, 1:2, 1:3, 1:4, and 1:5 for the former, and 1:1, 1:1.5, 1:2, and 1:4 for the latter.

Results

Most ventilators studied failed to consistently deliver the tidal volume set on the bellows control (Vt_{set}) when FGF was varied. Differences were also found between the various ventilators examined.

At 10 l/min FGF and Vt_{set} 1000 ml and I:E of 1:2, the Ohio 7000 delivered 1250 ml; a 25% discrepency; the ECAV 900 ml; the Dräger A-V and Narkomed 2 were accurate (Figs. 1 and 2).

At 4 l/min FGF, with the other settings unchanged, the volumes delivered were significantly less than Vt_{set} for all ventilators, except the Engström-Gambro. At FGF of 0.3 l/min, the difference was even more marked; the Dräger A-V delivered less than 700 ml; most the others delivered between 750 and 850 ml,

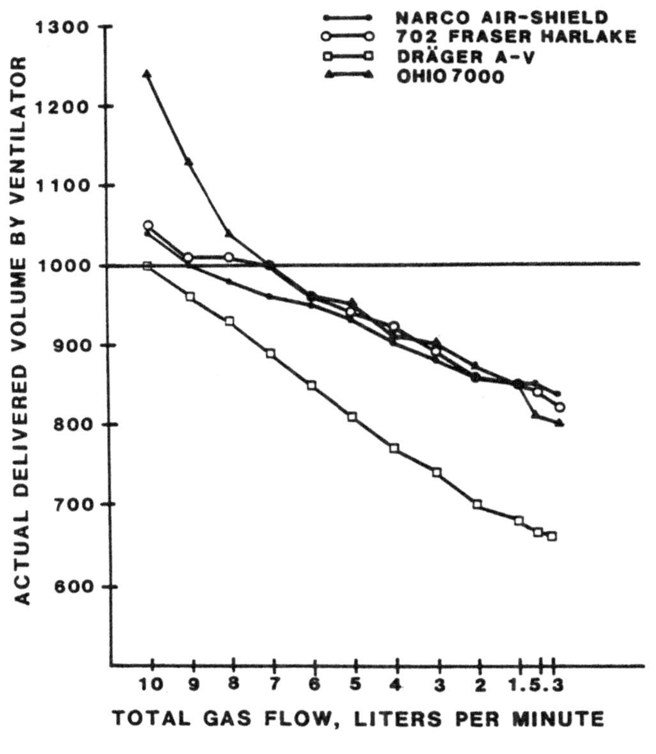

Fig. 1. Measured Vt matches its selected 1000 ml Vt only at fresh gas flow (FGF) rates greater than those used clinically; there is substantial disparity at low flows, especially for the Dräger A-V

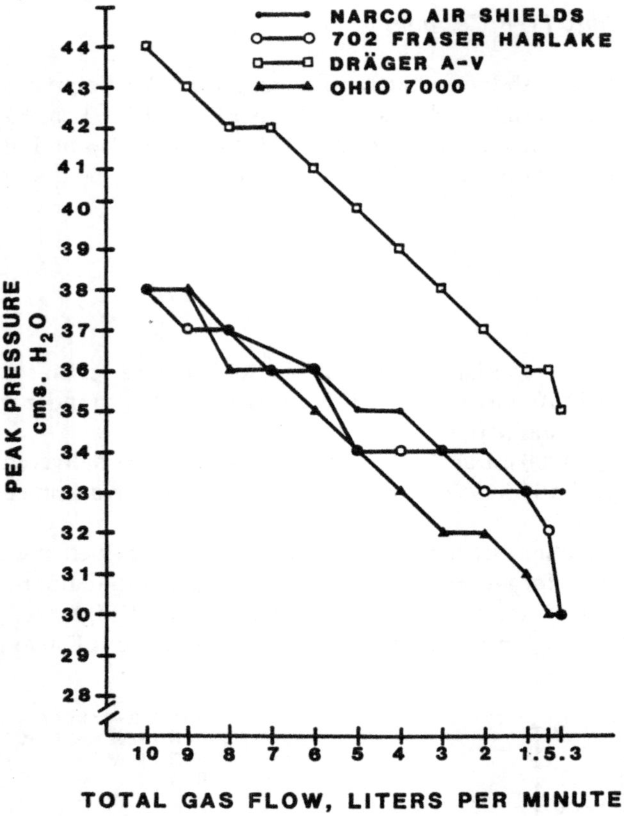

Fig. 2. For the same selected Vt, the peak airway pressure varies with FGF rate; the Dräger A-V consistently produced higher pressures than the others

the exception being the Engström-Gambro which delivered between 980 and 1130 ml (Fig. 3). Peak airway pressures changed too, proportionate to the FGF.

There was a significant correlation between changes in V and FGF (r = 0.998).

Preliminary observations with the Engström-Gambro experimental model anaesthesia ventilator revealed minimal variations of the measured to the indicated Vt with the 1:2 I:E ratio (Fig. 3). Similarly, the one single measurement at different FGF with the experimental ventilator Ohmeda 7800 showed minimal deviations from the expected dialed tidal volume, when FGF rates were less than 6 l/min (Fig. 4).

Measurements made with the Foregger 500 ECAV and the Narkomed 2A ventilators (Figs. 5 and 6) showed a similar decline as the earlier observatins with the 1:2 I:E ratio. Both apparatus, however, performed better as the denominator factor of the I:E ratio decreased. Although differences of Vt were lessened, as the expiratory phase increased throughout all FGF ratios, there was a reduction of Vt of about 20%.

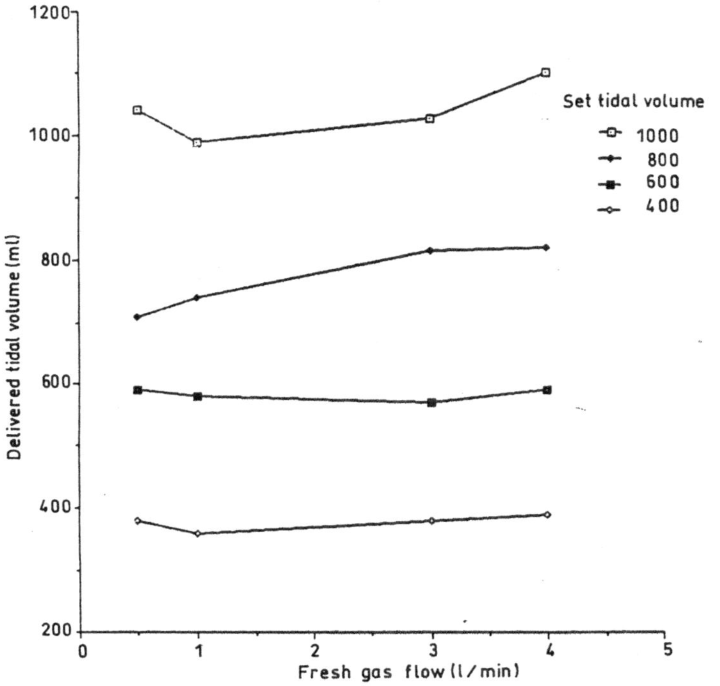

Fig. 3. Four different values for Vt were selected with the Engström-Gambro experimental ventilator. The volume delivered approximated that selected at FGF rates in common clinical use

Of interest was the parallel trend of all the curves produced by the Narkomed 2A ventilator with sigmoid-shaped tracings. Those curves resultant from the Foregger 500 ECAV ventilator, however, flattened at lower PGF rates, and there was no significant difference among the various I:E ratios. In contrast, at higher flow rates there were obvious differences among the various I:E ratios tested (Fig. 5).

Discussion

The measured VT of most of the ventilators studied was observed to be significantly different from that selected (Vt_{set}) by adjustment of the bellows displacement. The Engström-Gambro apparatus does not have an adjustable bellows housing; rather it delivers at Vt governed electronically by a valve which closes when the Vt_{set} has passed through a flowmeter.

Common sense suggests that ventilators should deliver the Vt_{set} which can be seen on the bellows housing, yet theory of ventilator design and the results of this study, are in conflict both with what would seem intuitively corrected and with each other.

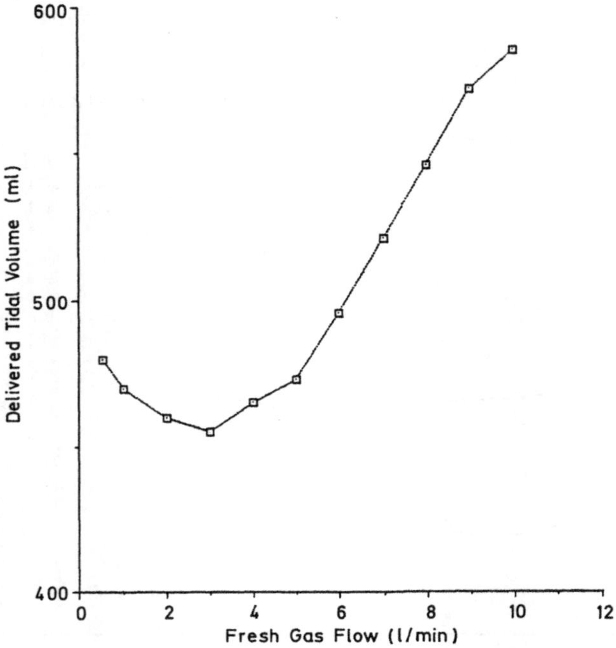

Fig. 4. A Vt of 500 ml was selected with the Ohmeda 7000 ventilator. At FGF rates of ≤6 l/min the variation of measured from selected volume was less than 8%

In a circle system, there are two uni-directional valves controlling the inspiratory and expiratory limbs, a bellows, a source of fresh gas, and a valve to exhaust excess gas from the systems [5]. The flow of fresh gas is constant and is not interrupted by the excursions of the bellows, so the amount of gas entering the circle system is that selected on the flowmeter controls; if this amount is greater than the patients need for oxygen or anaesthetic gases (minus the CO_2 produced) then an excess would accumulate unless discharged from the system [6]. The ventilators studied use a "spill-over" valve mechanism to vent the excess gases at the end of expiration, so that the volume of gas in the system from the end of one expiration to the beginning of the next inspiration consists of two parts; that exhaled by the patient (Vt) and the increment of fresh gas flow during the inspiration time. With a time ratio I:E of 1:2, one-third of the FGF contributes to the volume delivered to the patient. This can be expressed as follows:

$$Vt = Vt_{set} + \frac{FGF}{f\left(1 + \frac{1}{I/E}\right)} \quad \text{where f is frequency}$$

$$Vt = Vt_{set} + \frac{FGF}{f \cdot 3}$$

When Vt_{set} is 1000 ml, FGF is 10 l/min, and f is 10,

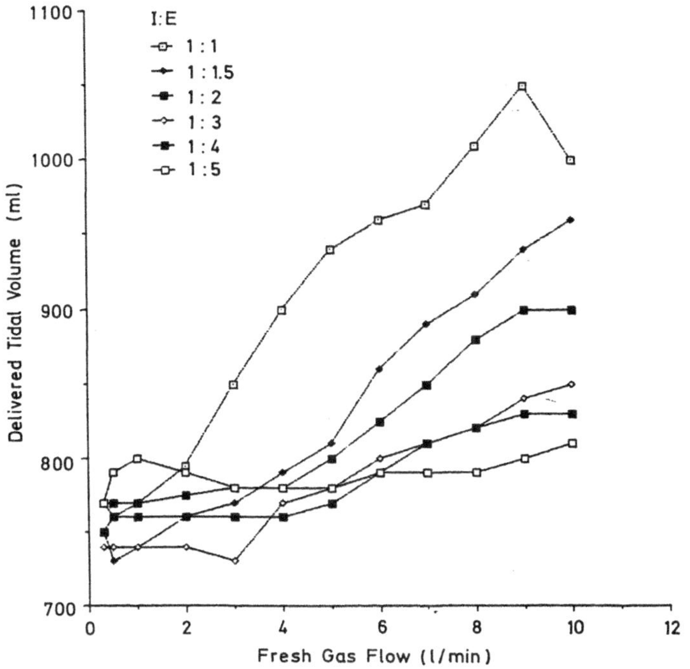

Fig. 5. The selected Vt was 1000 ml on the Foregger 500 ECAV ventilator. The delivered volume was less at various i:E ratios and FGF, specially at settings clinically used

$$Vt = 1000 + \frac{10{,}000}{30} = 1333$$

When FGF is changed to 1 l/min,

$$Vt = 1000 + \frac{1000}{30} = 1033, \text{ a decrease of 300 ml.}$$

Note however that the Vt is still greater than Vt set. What causes the Vt to be measured as less than calculated, or Vt_{set}, are the twin factors of compressibility and compliance [7].

When a gas is compressed, at constant temperature, its volume decreases; this is familiar as Boyles' Law [6]. For gas to flow in an anaesthesia system, a pressure differential is required, and when this is provided by the bellows the gas is compressed, its pressure increases and hence its volume diminishes.

Compliance may be thought of as the distensibility of structures in response to increased pressure [7]. Using plastic corrugated tubing in a circuit, and with the outlet at the common Y occluded, it was found that when 200 ml of gas were added to the system the pressure increased to 20 cms H_2O above ambient pressure. The compliance of the tubing is therefore 10 ml/cm H_2O. So, whatever Vt

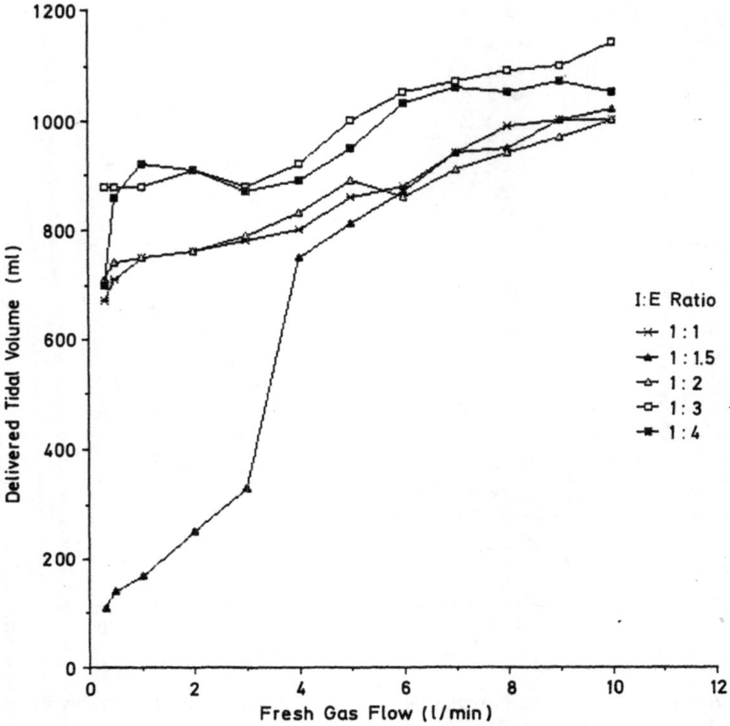

Fig. 6. On the Narkomed 2A ventilator, the Vt selected was 1000 ml. At clinically used settings of I:E of 1:2 and with FGF rates 4 l/min or less, delivered Vt was substantially less than the expected

may be set, 10 ml/cm H_2O pressure change will be needed by the circuit alone to satisfy compliance and compression requirements.

For example, FGF is 4 l/min, f is 10, V_{set} is 1000 ml, AWP is 25 cms H_2O.

$$Vt = 1000 + \frac{4000}{30} = 1133$$

Compliance requirements is $25 \cdot 10 = 250$.

Effective delivered Vt_{set} is 883.

If now we change the FGF to 1 l/min and AWP declines to 20 cms H_2O,

$$Vt = 1000 + \frac{1000}{30} = 1033$$

Compliance requirements is $20 \cdot 10 = 200$.

Effective delivered Vt_{set} is 833.

And if we now increase FGF to 8 l/min, and AWP increases to 30,

$$Vt = 1000 + \frac{8000}{30} = 1266$$

Compliance requirements is $30 \cdot 10 = 300$.

Effective delivered Vt_{set} is 966.

What this demonstrates is that the usual range of FGF used clinically will result is a Vt_{set} less than Vt_{set}, and that changes in FGF settings can alter the degree of ventilation of the patient [8].

For these reasons, the incorporation of an expiratory spirometer has been recommended [4, 5], in order to verify the actual Vt received by anaesthetized patients having their respiration controlled mechanically. Following peak pressure alone, as a guideline, may be misleading since this parameter may also be affected by changes of pulmonary and thoracic compliance, and does not appear to be a sensitive enough index of volume changes [5].

Conclusions

Anaesthesia ventilators differ one from another in the accuracy of their performance, and many in common use do not deliver the tidal volume selected. Moreover, the Vt delivered will vary with the FGF, and patients may be hypo- or hyperventilated unless capnography is used. In the usual clinical circumstances where FGF may be decreased following the initial period after induction, the patient may be hypoventilated. Conversely marked increases in FGF could subject the patient to high airway pressures.

The compliance of the breathing circuit must be considered in estimates of Vt and determinations of the settings to select; a spirometer measuring the actual Vt expired would be most helpful. It is also necessary to be familiar with the operating characteristics of equipment used in daily practice, especially the new.

References

1. Mushin WW, Rendell Baker L, Thompson PW, Maples OW (1980) Physical aspects of automatic ventilators: some applications of basic principles. In: Automatic ventilation of the lungs, 3rd edn. Blackwell Scientific, Oxford, pp 132–151
2. Cote CH, Petkan AH, Pyan JF, Welch JP (1983) Wasted ventilation measured in vitro with eight anesthetic circuits with and without in-line humidification. Anesthesiology 59:442–446
3. Ghani AG (1984) Fresh gas flow affects minute volume during mechanical ventilation. Anesth Analg 63:619–625
4. Aldrete JA, Castillo RA, Bradley ED (1986) Changes of fresh gas flow affect the tidal volume delivered by anesthesia ventilators. Anesth Analg 65:S4

5. Schreider P (1972) Anesthesia equipment, performance, classification and safety. Anaesthesiologie und Wiederbelebung. Springer, Berlin Heidelberg New York, pp 138–180
6. Stoelting RK, Miller RD (1984) Anesthetic equipment and breathing systems. In: Basics of anesthesia, 1st edn. Churchill Livingston, Edinburgh, pp 127–151
7. Stoelting RK, Miller RD (1984) Critical care medicine. In: Basics of anesthesia, 1st edn. Churchill Livingston, Edinburgh, pp 443–456
8. Twigley A, Chakrabarit, Whitwam JG (1987) A new ventilator for closed breathing systems. Anaesthesia 42:419–424

Limitations of Present Dosing Systems for Gases and Volatile Anaesthetics

H. Gilly

Introduction

The spreading application of anaesthesia techniques based on the administration of reduced fresh gas flows [4] places a greater emphasis on the demand for "precise" dosing of both volatile agents and anaesthetic gases. Occasionally, a more precise administration of anaesthetic gases is demanded for future anaesthesia delivery systems [1]. Due to their construction principles, vaporizers of conventional design are of only limited suitability for use with comparatively low gas flows, being optimized for flow-rates between 4 and 10 l/min [16]. Nevertheless, recent studies on units used clinically indicated a high incidence (36%) of significant deviations (> ±20%) between nominal and actual concentrations in this flow range as well [21]. As easy-to-use instruments enabling both a simple and a continuous measurement of anaesthetic concentrations have become available only recently, the differences between the concentrations set and actually delivered remained largely unrecognized during routine clinical work. In the following, the extent of these discrepancies will be described on the basis of our own investigations. Some of the underlying causes will be illuminated, and the accuracy limits of gas-dowing systems will be discussed.

Vaporizer Systems Currently in Use

Modern, calibrated vaporizers for volatile anaesthetics are placed outside the patient-circle system [5]. They are based on a "classical" construction [2] and, almost without exception, employ the principle of gas-flow partitioning: the concentration is adjusted by dividing the total gas flow, the smaller portion being fed through a vaporizing chamber where it becomes saturated with the anaesthetic agent [7]. The manner in which the gas-stream is partitioned limits the accuracy of the delivered concentration; in addition, it also determines the flow dependence and the magnitude of the concentration changes induced by fluctuations in back pressure due to ventilation [16]. Taking into consideration the physics involved as well as the construction itself (which both affect the concentration delivered by a vaporizer) then one cannot ignore the argument put forth by Rieke et al. [15] that even modern calibrated anaesthetic vaporizers have, as a result of their design, an inherent tendency to inaccurate concentration delivery.

Schreiber [16] formulated 12 criteria which an anaesthetic vaporizer should meet. The most important of these are the following:

- Independence from carrier gas flow
- Independence from changes in ambient temperature
- Temperature stability over time
- Insensitivity to changes in output pressure
- Safety
- Low resistance to flow

According to these criteria plenum vaporizers need to be designed in such a way to guarantee that the concentration delivered will be independent of temperature, gas composition, and the magnitude of the gas flow.

The Influence of Temperature

The anaesthetic concentration of the gas mixture in the vaporizing chamber depends on its partial pressure which can be read from the vapor pressure curve for the respective agent [18, 22]. The nonlinear temperature-dependent change in partial pressure is almost identical for halothane and isoflurane, which that of ethrane being largely similar. Assuming constant flow partitioning, the temperature characteristics would cause a nonlinear increase (decrease) in the output concentration as the temperature rises (falls). Therefore, so-called temperature-compensated vaporizers automatically adjust, by means of a (linear) compensation element, e.g. a bimetallic strip, that part of the gas stream which bypasses the vaporizing chamber [16, 17], thus counterbalancing the aforementioned nonlinear change. By this means the concentration delivered becomes largely independent of the temperature change as long as it varies between 20 and 30°C [17, 19]. However, if the temperature deviates by more than 10°C from room temperature, linear compensation is no longer sufficient to hold the concentration within 10% of the chosen value. For the sake of completeness, it has to be observed that the amount of anaesthetic vaporized becomes entirely unpredictable if the boiling point of the agent is approached. High gas flows together with high concentration settings may also cause marked discrepancies between the preset concentration and the delivered one. Since the device's thermal capacity is not sufficient to prevent a cooling of the vaporizer itself, under such circumstances, the delivered concentration falls so low [8] as to no longer meet relevant standards.

Effects of Gas Composition and Flow Rates

Stationary effects of variations in gas composition and/or flow rate on the vaporizer's concentration output depend, to a large extent, on the flow resistance of the vaporizing chamber and that of the bypass channel. It is essential that either a laminar or a turbulent flow profile exists in the gas flow paths and that at least the ratio of the flow resistance in both paths remains constant even if the velocity and/or the viscosity alters [16]. The concentration effects resulting from altered fresh gas composition and flow are quite well documented for the medium to high range of flow rates [12, 15].

We have investigated the range from 0.025 l/min to 20 l/min. In the following, the results for flow rates smaller than 2 l/min are given in detail. Figure 1 shows the setup used; measurements were done under the condition of a continuous gas flow without any back pressure on the vaporizer's output side. The flow rate and the composition of a binary gas mixture can be adjusted electronically using mass flowmeters (MKS 1258B) [11]. A commercially available infrared analyzer (IRINA, Dräger) [14] was used to quantify the anaesthetic concentration. Both the temperature of the unit and the ambient temperature, as well as the pressure drop across the vaporizer, were recorded. Tec4 and Vapor19 systems were tested, all of which had been checked within the past 6 months. Nominal concentrations were manually set with the vaporizer's control dial. For flow rates higher than 1 l/min, a new flow setting was maintained at least for 1 min before reading the measurements, for lower flow rates this period was increased up to 5 min. Air was exclusively used as carrier gas. All measured values were recalculated for an atmospheric pressure of 760 Torr and referred to a flow rate of 4 l/min for easier comparison of relative changes in concentration. Finally data were fitted using spline functions to obtain the smooth curves shown in the figures.

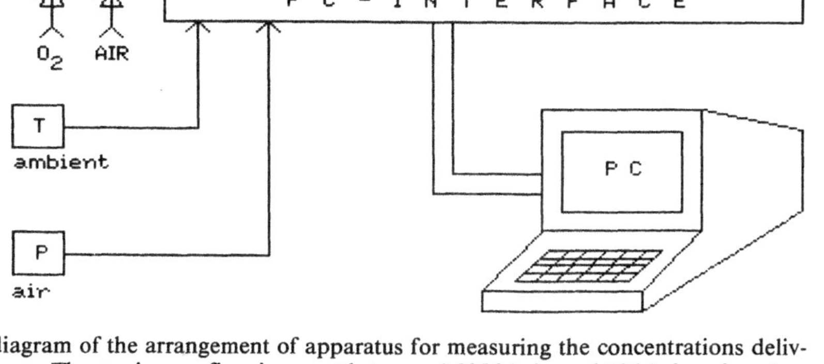

Fig. 1. Block diagram of the arrangement of apparatus for measuring the concentrations delivered by vaporizers. The carrier gas flow is preset between 0.025 l/min and 20 l/min using electronic flow controllers. The anaesthetic concentration is measured with an infrared analyzer. In addition, the pressure drop across the vaporizer, the ambient and the device temperatures are measured. All data are stored by a microcomputer system

Results of Concentration Measurements with Varying Gas Flow

When varying the carrier gas flow results as shown in Figs. 2–4 for Tec4 vaporizers and in Figs. 5–7 for the Vapor19 were obtained.

Under the test conditions described above, the concentration delivered by the Fluotec4 vaporizer varied between +5% and −8% over the flow range from 0.2 to 8 l/min (Fig. 2); values given are referred to those obtained at a flow rate of 4 l/min. At the lowest dial setting (0.25 vol%), however, larger deviations became apparent: thus, for flow rates below 0.2 l/min the concentration delivered decreased, the relative change being about −15%. Figure 2 also illustrates the relative deviations in the high flow range typical of the Tec4 series: at distinct concentrations rather gross differences become obvious when comparing the high and medium flow ranges.

Using the highest dial setting (5 vol%) of the Enfluratec4 vaporizer (Fig. 3) the reading differed notably from the nominal value: the delivered concentration was increased by 10%, the extent of this deviation remaining rather constant for flow rates higher than 0.2 l/min. Even larger positive differences were found for low concentration settings at flow rates less than 0.2 l/min. Using the dial settings of 1 and 2 vol% the delivered concentrations varied only slightly (+4% to −5%) when changing the carrier gas flow.

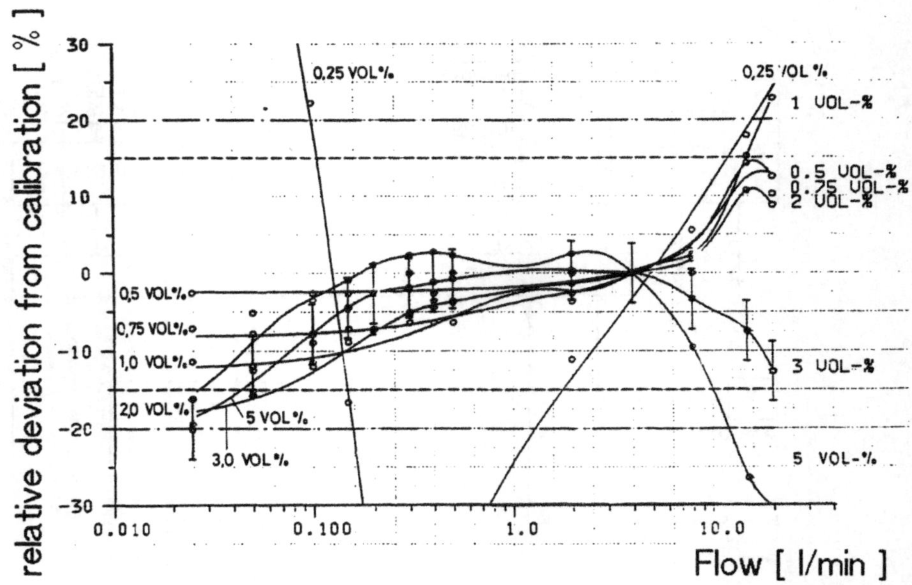

Fig. 2. Relative changes in the halothane concentration delivered by the Fluotec4 vaporizer, as a function of the carrier gas flow (0.025 to 20 l/min). Carrier gas: air. Room temperature. The values measured at 0.025, 0.05, 0.1, 0.2, 0.3, 0.4, 0.5, 2, 4, 8, 15 and 20 l/min are expressed in relation to the concentrations delivered at 4 l/min. Between 0.5 and ca. 6 l/min, the concentrations delivered vary about ±5% for all calibrated dial settings except the lowest

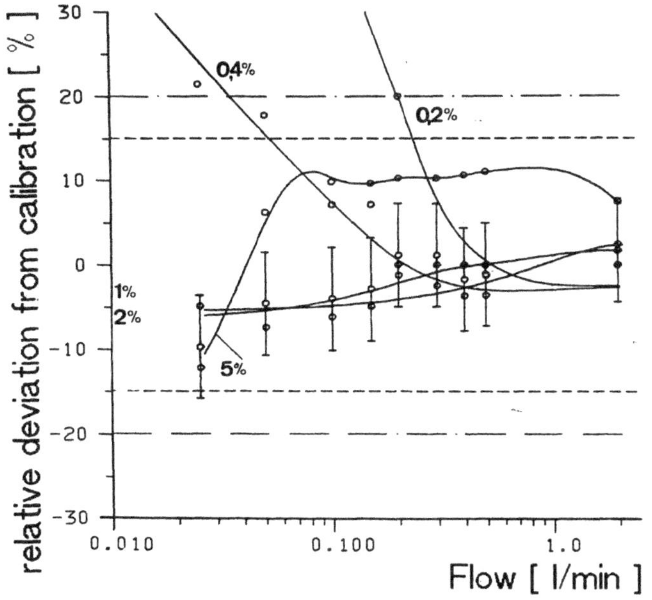

Fig. 3. Relative changes in the enflurane concentration delivered by the Enfluratec4 vaporizer, as a function of the gas flow (0.025 l/min to 2 l/min). Measurement conditions as in Fig. 2. At the lowest dial setting a reduction in flow below 0.5 l/min increases the delivered concentration markedly. For the highest concentration setting, a relative change of ±10% is observed which is practically independent of flow ranges from 2 l/min down to 0.05 l/min. Marked reduction in delivered concentrations below 0.05 l/min. This phenomenon is not observed at the high or mid-range concentration settings

The isoflurane concentrations delivered at various flow rates are shown in Fig. 4. When using a high flow and when setting a high concentration (3 and 5 vol%) the measured concentrations differed by up to 1.2 vol% from the expected ones. For the lowest dial setting (0.25 vol%), a flow dependence similar to that of the halothane vaporizer was observed. At all other nominal concentration settings, the discrepancies did not exceed +3% and −8% respectively when measurements were made within the flow range specified by the manufacturer. Further reductions of the flow rate resulted in larger relative deviations increasing up to a maximum of −15%.

Figure 5 shows the changes in the output concentration (relative values) as a function of the flow rate (0.025 l/min to 20 l/min) for the Vapor19 halothane unit. Within the range specified by the manufacturer (0.5–15 l/min) deviations do not exceed ±3%. At the lowest dial setting the relative deviations increase at flows less than 0.5 l/min, differing by 25% at a flow rate of 0.1 l/min. At all other calibrated dial settings, the maximal decrease within the specified range was 8%, increasing to about 15% at 0.025 l/min when selecting 2, 3 and 4 vol%.

Figure 6 depicts the relative deviations in output concentration of a Vapor19 enflurane vaporizer for flows ranging from 0.025 to 2 l/min. The concentration

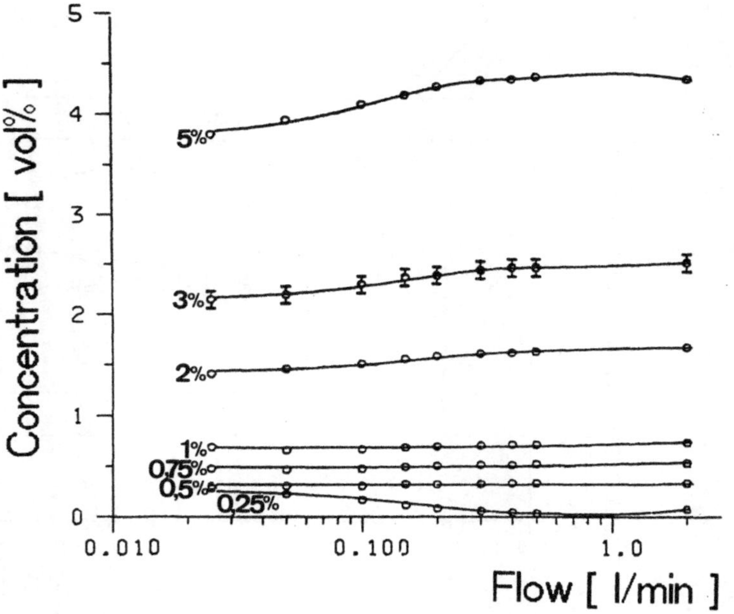

Fig. 4. Concentration output of an Isotec4 vaporizer as a function of gas flow (0.025 to 2 l/min). Measurement conditions as in Fig. 2. Pronounced changes in the isoflurane concentration delivered are found outside the manufacturer-specified flow range at dial settings of 3 and 5 vol%. At 0.25 vol%, noticeable though not clinically relevant deviations from the nominal value are found within the manufacturer-specified flow range

delivered is more or less independent of flow between 0.2 and 2 l/min: the differences lie within about ±2% of the respective concentration measured at 4 l/min. Similar to the halothane vaporizer (Fig. 5) a reduction of the flow down to 0.025 l/min results in larger differences (about −4% to −12%), excluding the lowest concentration setting.

In Fig. 7 the flow dependence of the Vapor19 isoflurane is shown. At a setting of 5 vol% and a flow rate of 1 l/min the concentration delivered deviated by 0.45 vol% from the nominal value. This difference increased to almost −1 vol% at a flow of 0.025 l/min. Somewhat smaller relative deviations were found for concentration settings below 2 vol%.

These results may be summarized as follows: Within the flow range studied, the nominal and actually delivered anaesthetic concentrations vary, the differences being largely independent of the type of unit. The Tec4 systems consistently exhibited larger absolute deviations. Presumably the most likely explanation for these discrepancies is the fact that these vaporizers are calibrated by the manufacturer using oxygen as the carrier gas [17]. In contrast, our measurements were made exclusively using air. The different viscosities of air and pure oxygen are known to cause changes in concentration output of up to 25% [12], the actual difference also depending on the flow rate. Thus, in comparison with the Va-

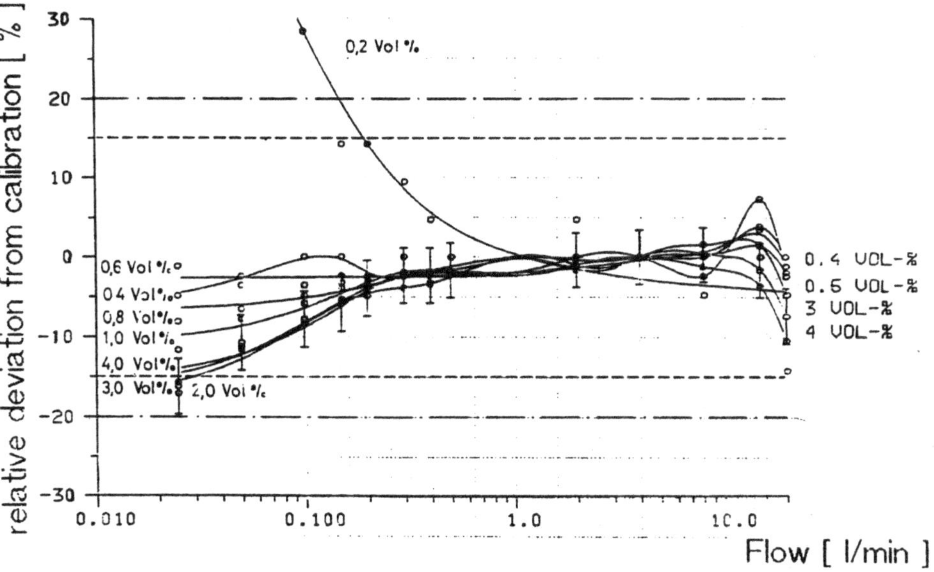

Fig. 5. Relative changes in the halothane concentrations delivered by a Vapor19 unit. Measurement conditions as in Fig. 3. With the exception of the lowest concentration settings (0.2 and 0.4 vol% resp.), deviations are within about ±3% for flow rates specified by the manufacturer (0.5 to 15 l/min). At the higher dial settings, the halothane concentration delivered is markedly reduced at flow rates below 0.2 l/min

por19 units, the larger "inaccuracy" that we observed in the Tec4 vaporizers could possibly be due to the different calibration conditions.

Both systems investigated showed flow-induced deviations from nominal concentration output: the range of these changes was limited to about ±10% as long as the flow rate remained within the range specified by the manufacturer and as long as the preset concentration was higher than 0.4 vol%. Outside the manufacturer-specified range, flow-induced relative deviations of up to 15% were observed and, in the case of the Fluotec4, the relative decrease in the concentration delivery was almost −20% at 0.025 l/min. When using very low flows at the lowest dial setting, positive deviations exceeding the values stated above were found.

Prospects of Improvements Enhancing the Accuracy of Vaporizers

The influence especially of gas composition and flow rate on the output concentration can be reduced by abandoning the principle of the plenum vaporizer, i.e. saturation of the gas mixture in a relatively large vaporizing chamber and subsequent dilution employing mechanical flow partition. The alternatives which may be considered are (i) the direct vaporization of an exact amount of the volatile anaesthetic in liquid form, and (ii) the (electronically controlled) addition of a

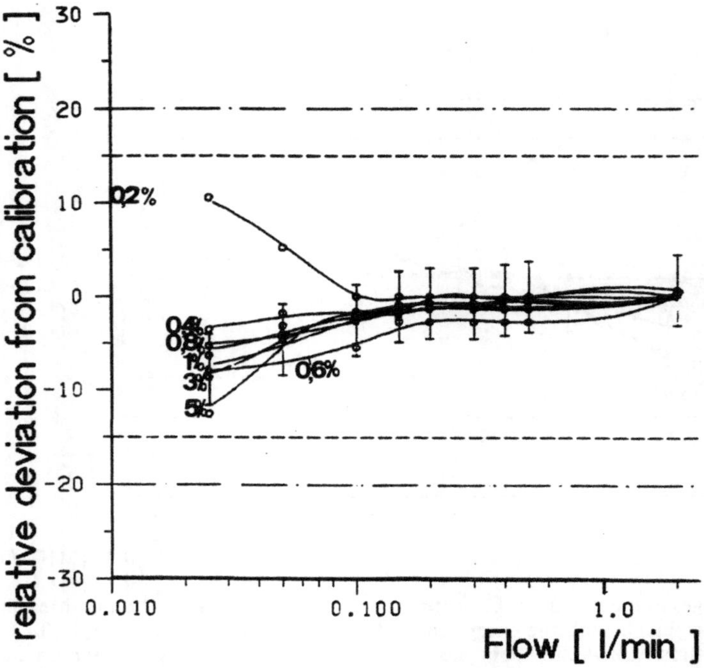

Fig. 6. Relative changes in the enflurane concentration output of the Vapor19 unit, as a function of the gas flow varying from 0.025 l/min to 2 l/min. Flow-induced changes in concentration worth mentioning are found only at concentration settings of 0.2 and 5 vol% for flow rates between 0.025 and 0.1 l/min

defined gas volume of known saturation to the fresh gas. This last method was suggested by Hahn et al. [10] and, for halothane, flow-independent concentration delivery for settings from 1 to 4 vol% could be achieved in the range between 2 and 16 l/min. The observed deviations were, at most, ±0.1 vol%, which corresponds to 3%–10% of the value set. With the oxygen content between 20% and 50% the influence of the carrier gas composition on delivered concentration was up to ±0.2 vol% and thus somewhat higher than the flow-induced change but still considerably lower than the carrier gas-induced changes seen in vaporizers of conventional design.

An electronically controlled vaporizer designed by Engström for the latest generation of anaesthesia breathing systems mixes the fresh gas flow with a defined volume of virtually 100% saturated anaesthetic gas produced through an increase in temperature. The essential differences between this vaporizer and that of Hahn et al. [10] is that fresh gas is not passed through the vaporizing chamber; subsequently a dependence on flow or gas composition is not to be expected. Since, as far as we know, no published studies on the new Engström vaporizer are available, and since we have not yet been able to make our own measurements, we have to refer here to the specifications stated by the manufacturer.

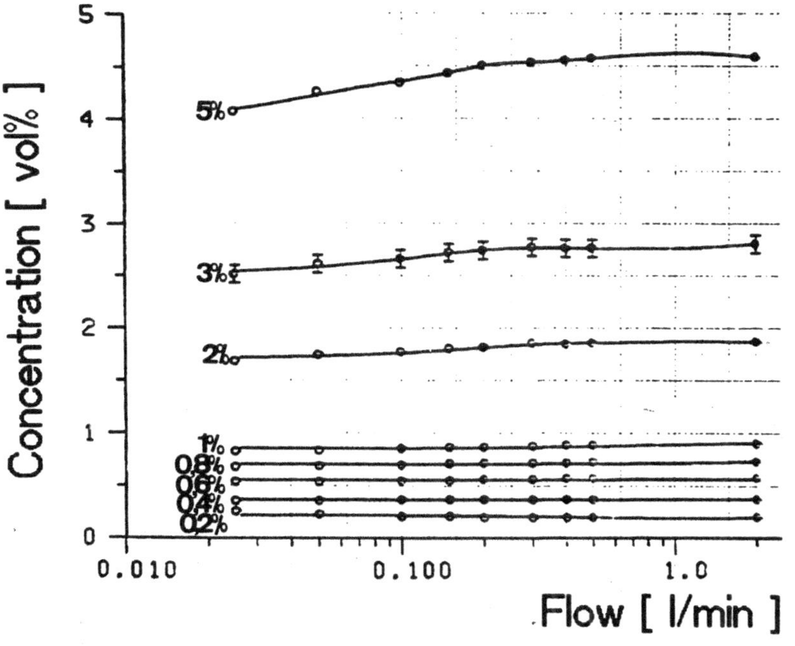

Fig. 7. Isoflurane concentrations delivered by a Vapor19 isoflurane vaporizer. Measurement conditions as in Fig. 2. Nominal and actually measured concentrations differed by up to −10%, and are rather independent of the preset concentration. At decreased flow rates the concentration delivered at the 3 and 5 vol% settings was reduced by almost 20%

In the vaporizers made by Siemens-Elema (types 950, 951, 952), the liquid anaesthetic is directly injected into the fresh gas stream. The pressure difference established across a stenosis determines the amount of liquid injected. (Given the pressure, temperature, and the corresponding physicochemical constants, the resultant gas volume of the volatile anaesthetic can be calculated by using the equation of state for real gases.) With this type of vaporizer the accuracy of the concentration delivered primarily depends on keeping the fresh gas flow velocity across the stenosis constant. According to our own experience [9] the gas flow was subject to minor fluctuations in the older models. Therefore, variations in the respiratory minute volume also affected the output resulting in noticeable concentration changes. A significant reduction in the flow dependence has since been achieved by changes in design. (A. Obermayer 1987, personal communication.)

A new vaporizer system was recently disclosed by the Dräger company. A digitally controlled microcylinder pump adds defined amounts of the liquid anaesthetic to the fresh gas. This technique seems to be a promising approach to achieving precise concentration delivery. (See Wallroth, this volume, for further details.)

In comparison with conventional plenum vaporizers, all the electronic dosing systems for volatile anaesthetics mentioned above have two additional advantages: (i) they may be remotely controlled, and (ii) the range of concentrations that can be set is larger. This allows a more rapid build up of the anaesthetic concentration in the patient-circle in minimal-flow anaesthesia and in the closed-circuit system. Up to now special plenum vaporizers with an increased range had to be used for this purpose.

Accuracy of Gas Dosing

In order to achieve a most accurate dosing of the volatile anaesthetic agents both the concentration and the amount of fresh gas must be precisely set (and measured). Pressure compensated rotameters of conventional design have a relatively limited range of control (typically 0.5 to 20 l/min, depending on the type of unit used) with a rather coarse resolution limit. The accuracy is quoted as ca. ±3% of maximum scale value, which corresponds to a flow adjustment of ca. 10% at the mid-range [16]. In order to achieve fairly accurate oxygen flow settings at low flow rates, therefore, an additional rotameter with a reduced measuring span is needed.

Flow-measuring techniques based on other physical principles such as the hot wire anemometer, the differential pressure transducer with linear flow resistances (Fleisch tube pneumotachometer) etc. show inaccuracies in the order of ±10%. By using the vortex method a measurement accuracy of about ±3% is obtained, however, the measurement range is limited to 2–12 l/min [20]. This comparatively high accuracy seems to be exceeded by new sensors based on the ultrasound Doppler technique [3]. An even somewhat better performance is attained in the so-called mass-flow controllers, in which the measurement is based on the amount of energy needed to maintain a particular thermal profile along the measuring channel [11]: for a single component gas an error of less than 1% of maximum scale value is easily obtainable. The major disadvantage of mass-flow controllers, however, is that binary and tertiary gas mixture can be set with the accuracy stated above only if the concentrations of all different gas components are known exactly. (These flow controllers seem to have been used up to now exclusively for the precise adjustment of exact anaesthetic concentrations when testing the accuracy of gas analyzers [23].)

Despite some considerable improvements in dosing accuracy achieved with time-controlled pneumatic valves [6, 13] as well as electronically feedback controlled high-pressure servo valves (HPSV; Dräger company), no satisfactory solution has yet been found to the problem of exact dosing of breathing gases. According to Wallroth (this volume) flow or volume settings with an approximate error less than 5% could be achieved by using gastight dosing cylinders displacing the gas volume at constant pressure (the scoop principle); only recently does the introduction of rolling seal membranes for reduction of the gas leakages seem to provide a technique that allows for the exact determination and dosing of gas volumes even with varying gas composition.

The Fields of Application for Conventional Anaesthetic Gas-Dosing Systems

In view of the limited accuracy of current anaesthetic gas-dosing systems described above, we must ask whether anaesthetic vaporizers presently available are too inaccurate, and particularly whether enhanced accuracy of the concentration delivered is urgently indicated. Besides the basic expectation that manufacturers will continually keep pace with technological advances, both improving their products' reliability and observing relevant standards, there are at least two goals, by no means compatible, to be taken into consideration: the demand for new dosing systems (and anaesthesia techniques) enabling exact balancing of the amount of anaesthetic gas delivered and consumed ("quantitative anaesthesia"), and the requirement for anaesthesia techniques which are simple to carry out in routine clinical work. In the latter case a limited accuracy in anaesthetic dosing such as that provided by the devices of the latest generation will suffice. This does not mean, however, that any differences between nominal and actually delivered concentration can be considered unimportant or even disregarded altogether. Since semi-closed breathing systems supply the volatile agent "in excess" even when slightly reduced fresh gas flows re-used, the individual variations in anaesthetic consumption are of minor importance in comparison to anaesthesia performed using a closed-circuit system. A more general application of minimal-flow and closed-circuit techniques - presently restricted to a few experts only - is based on the condition that the concentration of breathing and anaesthetic gases in the patient-circle system is closely controlled and that the vaporizers provide both higher accuracy and a wider concentration range than the ones presently in use. This implies the abandoning of mechanical dosing systems such as exemplified by the plenum type vaporizers and rotameters with floating bobbins in favor of more precise electronic systems.

Summary

The majority of calibrated dosing devices nowadays used clinically for volatile agents and anaesthetic gases are based on "classical" construction principles which have remained largely unchanged within the past decases. With these conventional plenum vaporizers, the concentration of a volatile anaesthetic can be set with a fair degree of accuracy within the usual clinical range (0.2–2 MAC) and for intermediate flow rates (2–8 l/min). When using high fresh gas flows, these vaporizers show deviations, at times drastic, from expected concentrations. At reduced gas flows, such as those used in minimal-flow and closed-circuit systems, deviations of $\pm 20\%$ of the nominal value must be expected in the intermediate concentration range. If anaesthetic gas composition, flow velocity and temperature change simultaneously, the resultant difference between set and delivered values can no longer be quantified precisely. When setting a flow in the conventional way by means of a floating-body rotameter, exact dosing of the volatile anaesthetic will hardly be feasible in the low-flow range (<2 l/min). Vaporizers and gas-dosing systems presently available should only be used in low-flow or closed-circuit systems if, as a minimum requirement, continuous monitoring of either the inspiratory or expiratory anaesthetic concentration is provided. For this kind of anaesthesia technique the technical basis should now be available to achieve the highly desired precision of dosing, even at higher concentrations (6–8 vol%), employing liquid injection vaporizers. These technical improvements will on the one hand allow truly quantitative dosing of anaesthetics, and on the other hand provide the preconditions for semiautomatic and remote control of the concentration of anaesthetic gases in the circuit system.

References

1. Bergman H (1986) Das Narkosegerät in Gegenwart und Zukunft aus der Sicht des Klinikers. Anaesthesist 35:587–594
2. Boulton TB (1985) The origins of the plenum principle of vaporization. Surv Anesthesiol 29:201–204
3. Buess C, PEach P, Guggenbuehl W, Koller EA (1986) A pulsed diagonal beam ultrasonic air flowmeter. Appl Physiol 61 (3):1195–1199
4. Conway CM (1984) Closed and low-flow systems. Acta Anaesthesiol Belg 4:257–263
5. Conway CM (1985) Anaesthetic breathing systems. Br J Anaesth 57:649–657
6. Cooper JB, Newbower RS, Moore JW, Trautman ED (1986) A new anesthesia delivery system. Anesthesiology 49:310–318
7. Dorsch JA, Dorsch SE (eds) (1984) Understanding anesthesia equipment. Construction, care and complications, chap 4, 2nd edn. Williams and Wilkins, Baltimore, pp 77–135
8. Fitzal S, Gilly H, Steinbereithner K (1986) Do modern plenum vaporizers provide accurate anesthetic mixtures irrespective of gas flow? Anesthesiology 65:A168
9. Gilly H, Steinbereithner K, Tobolik G, Watzek C (1986) Sicherheit durch Überwachung von Narkosegaskonzentrationen volatiler Anästhetika. In: List WF, Bergmann H, Schalk HV (eds) Anästhesie im kleinen und mittleren Krankenhaus. Anästhesiologie und Intensivmedizin, vol 192. Springer, Berlin Heidelberg New York, pp 145–159
10. Hahn CEW, Palayiwa E, Sugg BR, Lindsay-Scott D (1986) A microprocessor-controlled anaesthetic vaporizer. Br J Anaesth 58:1161–1166
11. MKS-Flowmeter instruction manual: MKS mass flowmeter and flowcontroller manual 195-109 500-B-2/86. MKS Instruments, Andover MA 01803
12. Palayiwa E, Sanderson MH, Hahn CEW (1983) Effects of carrier gas composition on the output of six anaesthetic vaporizers. Br J Anaesth 55:1025–1038
13. Palayiwa E, Hahn CEW, Sugg BR, Lindsay-Scott D, Tyrrell PJ (1986) A microprocessor-controlled gas mixing device. Br J Anaesth 58:1041–1047
14. Pockrand I (1985) Optische Gasanalyse in der Medizin. Technisches Messen 52:247–252
15. Rieke H, Raffauf EM, Züchner K, Hauschke D, Sonntag H (1985) Genauigkeit von Halothanverdampfern in Abhängigkeit von Temperatur, Trägergaszusammensetzung und Gasflußrate. Anaesthesist 34:522–527
16. Schreiber P (1972) Anaesthesia equipment. Performance, classification and safety. Springer, Berlin Heidelberg New York, pp 30–72 (Anästhesiologie und Wiederbelebung vol 59)
17. Tec4 continuous flow vaporizer operators manual. September 1984 CY 548 Ohmeda, Keighley, Yorkshire
18. Terrell RC (1984) Physical and chemical properties of anaesthetic agents. Br J Anaesth 56:3S–7S
19. Vapor19 Betriebsanleitung 5127.0 3. Ausgabe Februar 1979. Drägerwerk AG, Lübeck
20. Westenskow DR, Tucker SM (1981) Evaluation of a ventilation monitor. Crit Care Med 9:64–66

21. Whitcher C, Parsons D, Ream AK, Rubsamen D, Scott J, Champeau M, Sterman W (1986) Anesthetic safety and cost effective monitoring. Scientific Exhibit, ASA Meeting 1986, pp 1–10
22. White DC (1985) Vaporization and vaporizers. Br J Anaesth 57:658–671
23. Zbinden AM, Westenskow D, Thomson DA, Funk B, Maertens J (1986) A laboratory investigation of two new portable gas analysers. Int J Clin Monit Comput 2:151–161

Technical Realisation of Quantitative Metering and Ventilation

C. F. Wallroth, R. Jaklitsch, and H. A. Wied

Introduction

An attempt will be made to specify and define quantitative delivery systems for anaesthetic gases, liquids and vapours, and respiratory volumes. Accuracy limits of existing and forthcoming metering and ventilation systems will be established to select the most effective system for quantitative anaesthesia. Ways of its technical realisation will be presented.

Accuracy Requirements

We define quantitative anaesthesia as the accurate knowledge of quanta of gas, liquids and vapours, and respiratory volumes delivered to and from the patient. For the quantitative metering we need to know the quantum (lat. amount) delivered over a defined period of time. Time can be measured very accurately; however, the measurement of a quantum of gas, liquid, vapour or volume is much more difficult.

Naturally some questions with respect to the required accuracy arise: How accurate should metering be? Is an accuracy of ±30%, ±15% or ±5% sufficient for quantitative anaesthesia? The question about required accuracy will have to be answered by the clinician's need for information and the usefulness of this information, e.g. for diagnosis and therapy of an individual patient.

An example of a physiological parameter with a low variability is the change in metabolic oxygen uptake with temperature: a drop of 1° in core temperature results in a 10% decrease of metabolic oxygen uptake. To obtain quantitative information about patient uptake one would measure concentration and flow in the inspiratory and expiratory path. Then the uptake can be calculated from the relatively small difference of two numbers. For example:

Inspired oxygen concentration

$$FiO_2 = 40\% \pm 3\%$$

Expired oxygen concentration

$$FeO_2 = 36\% \pm 3\%$$

Average output (minute volume)

$$\dot{V}_{mv} = 6\ \text{l/min} \pm 0.5\ \text{l/min}$$

Under the assumption of uncorrelated measurements this will result in the following oxygen uptake:

$$\dot{V}_{O_2} = (FiO_2 - FeO_2)\,\dot{V}_{mv} = 240\ \text{ml/min} \pm 255\ \text{ml/min} \quad (1)$$

This means that despite reasonably accurate measurements, the signal-to-noise ratio becomes approximately 1 and the quantitative uptake information is masked by the measurement error.

An overall accuracy of ±30% or more is probably unacceptable for quantitative anaesthesia; ±15% appears marginal and ±5% or better is desirable. But are they obtainable with available or forthcoming equipment? Before establishing guidelines a survey of existing equipment is warranted. The accuracy limits of present-day equipment are summarized in Table 1.

For the delivery of gases a needle valve is adjusted manually until the flowmeter shows the desired flow rate. An accuracy of ±10% of the actual reading can be achieved at flow rates greater 300 ml/min. However, this accuracy applies to single-parameter variation only. Ambient pressure and back pressure, ambient temperature and gas temperature, gas composition and gas condition (e.g. lack of dryness and cleanness) bias the reading. For the delivery of the anaesthetic agent, a vaporizer is set to the desired rate of concentration whereby the liquid agent is exposed to the fresh gas flow in a controlled fashion. Present-day accuracy is ±15% of the actual reading for single-parameter variation when compared with the manufacturer's calibration condition.

Tidal volumes are delivered by means of a compliant bag or bellows. The product of tidal volume and breathing frequency forms the minute volume. Therefore, the accuracy of both parameters determines the accuracy of the delivered minute volume according to the sum of squares law. Since the accuracies of minute volume and frequency are about ±15% each, the accuracy of the minute volume is about ±20% of actual reading.

Table 1. Accuracy of present-day equipment[a]

Device	Variable	Accuracy
Flow meter	Flow	±10%
Vaporizer	Concentration	±15%
Ventilator	Tidal volume, respiration rate	±15%

[a] Accuracy applies to single parameter variation only

Such equipment for the delivery of gas, liquid and minute volume could be calibrated to a high degree of accuracy over the range of application and environmental conditions on an individual component basis. For example a flow meter tube and the associated needle valve mounted on the actual anaesthesia machine could be calibrated to an accuracy of about ±3% of actual reading. A soap bubble film flow meter could be used for this calibration over the range of expected temperatures, pressures, ventilator settings, and other variables. All this would result in a huge amount of multiparameter measurements and correction tables valid for only one individual set of equipment. These correction tables are extremely difficult to handle by the user. Thus, individual calibration of present-day equipment to improve the accuracy beyond technically feasible values is impractical and uneconomical.

The applied technologies are based on mechanical and pneumatic principles. They utilize analogue techniques for gas delivery (e.g. needle valves), gas metering (e.g. flow meter tubes), vaporization (e.g. plenum vaporizer or bubble through vaporizer), and ventilation (e.g. volume pumps with compliant compartments). These analogue technologies do not allow data logging, which is absolutely necessary for quantitative anaesthesia. Especially if the metering has to be changed frequently, manual recording and subsequent calculation of patient uptake is tedious and time-consuming. Such tasks can easily be performed by a computer. With respect to quantitative anaesthesia the limited accuracy of mechanical-pneumatic-analogue devices in combination with the lack of data logging lead to a technological blind alley. How can we get out of it?

Alterantive delivery means have to be analyzed with view to the following design objectives: safety and reliability; accuracy; reproducibility or precision; stability; maintainability; data logging capability; and remote control feasibility.

Gas Delivery System

First let us take a look at anaesthetic and respiratory gas delivery. What are the objectives when designing such a delivery system?

One of the most important objectives certainly is safety and reliability (see Table 2). With respect to accurate gas metering for the purpose of quantification

Table 2. Design objectives for quantitative gas metering

- Safety and reliability	O_2-N_2O ratio control; O_2 supply failure alarm; N_2O cut-off safeguard against inadvertent adjustment; failure detection capability; high reliability; fail-safe design
- Overall accuracy	±5%
- Reproducibility	±2%
- Stability	$< \pm 1\%$/month
- Response time	$T_{90} \leq 5$ s
- Dynamic range	50 to 12000 ml/min (1:240)
- Data logging	
- Remote control	

one would expect an accuracy of less than ±5% of the actual value, independent of ambient temperature and back pressure. Flow rates should be reproducible (±2% of the actual value) and the rate of drift should be less than ±1% per month. Low flow anaesthesia requires fresh gas delivery systems which provide reproducible fresh gas flows as low as the uptake rate for adult and pediatric patients. Therefore, flow rates should start at about 0.05 l/min. The range of application should cover open, semi-open, semi-closed, and closed systems. This requires flow rates of 0.05–12 l/min, which resembles a dynamic range of 1-240.

Now three methods of anaesthetic gas delivery will be presented for comparison:

- Conventional metering with a calibrated flow meter as sensor and a needle valve as actuator;
- Analogue metering using a servo loop;
- Discrete digital metering employing a volumetric dosage principle.

Conventional metering (Fig. 1) with a flow meter and a needle valve involves an operator (i.e. the anaesthetist), a measuring device (i.e. the flow meter), and an actuator (i.e. the needle valve). The anaesthetist adjusts the flow rate by changing the position of the needle valve and at the same time observes the float in the flow meter. The position of the float with respect to a measuring scale provides the measured value. The anaesthetist compares this value with the one he desires, i.e. the setpoint. If he is not satisfied with the set flow rate, he takes further action and adjusts the flow rate again. This is a process, where the anaesthetist is part of a loop, which is generally called a feedback loop or closed loop (in German, *Regelkreis*). When anaesthetic gases are metered in such a way, it is difficult to quantify and account for how much is actually delivered, because total accuracy is a function of flow meter accuracy, stability of the needle valve and reliability of the operator.

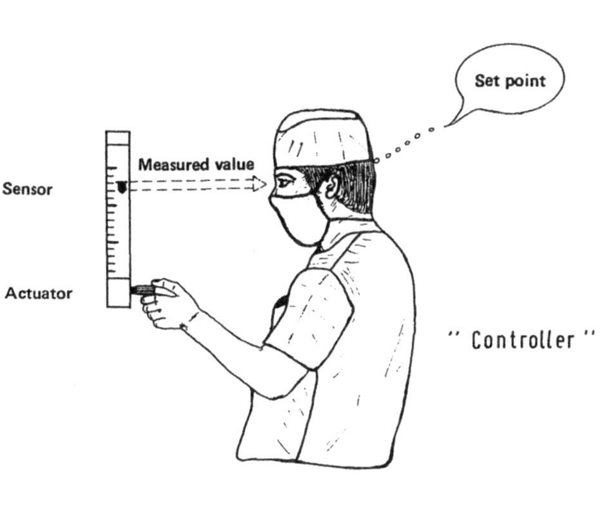

Fig. 1. Metering controlled by anaesthetist

Therefore, the logical consequence of this would be to automate this process of adjusting the flow rate employing servo-loop control (Fig. 2), thereby making quantification feasible. For this purpose a flow sensor transduces flow rate into an electronic signal, i.e. the measured value, which is fed back to the controller and compared with the desired flow rate. A difference between the measured value and the desired value causes the controller to adjust the actuator, until the desired value is practically reached. It is obvious, that the accuracy of metering is limited by the accuracy of measurement of the flow rate. Nowadays flow sensors have an accuracy of ±8% to ±15%, which also limits the accuracy of metering. This is the major disadvantage of analogue metering devices using a servo-loop. In addition, the system is not inherently fail-safe, since no redundant monitoring of the flow rate is done. However, this method provides an electronic signal necessary for automatic data logging. The system is not able to discriminate between set value and actual value, because the actuator follows the measured value only; set value and measured value become identical.

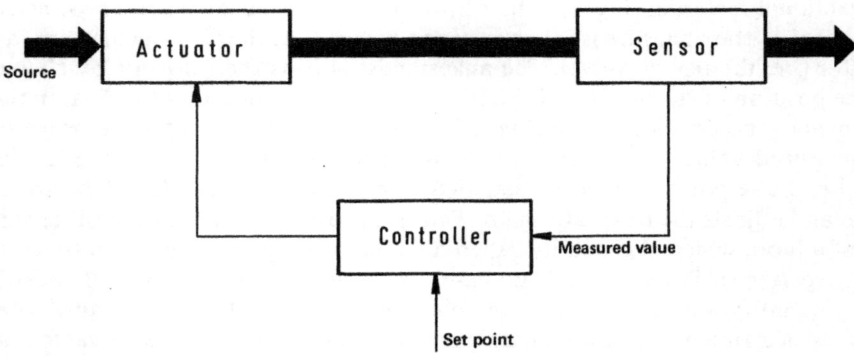

Fig. 2. Analogue metering with servo-loop

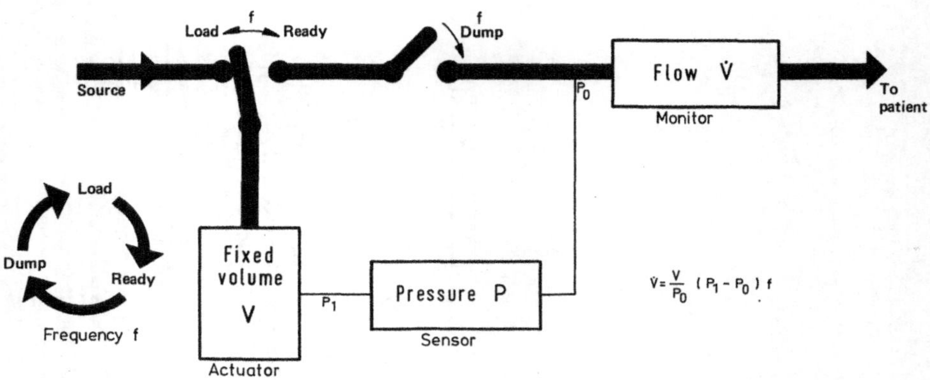

Fig. 3. Digitally controlled volumetric gas metering

The above-mentioned design objectives, including inherently fail-safe design, are met with the principle of discrete volumetric dosage. A pictorial description is found in the measuring cup principle, where well-defined discrete amounts (quanta) are added until the desired value is reached (Fig. 3). A fixed volume, a compartment of a known quantum, is filled at a preset pressure P_1 by opening a valve, which then switches to the 'ready' position. The fixed volume is thereafter released to a second preset pressure P_0 by switching a second valve. The operation consists of three strokes:

1st stroke: Load stroke
2nd stroke: Ready stroke
3rd stroke: Dump stroke

The pressure difference $(P_1 - P_0)$ and the frequency f describe the volume delivered over time:

$$\dot{V} = \frac{V}{P_0}(P_1 - P_0)f \tag{2}$$

System function is monitored by an independent flow monitor, e.g. a hot wire anemometer, which is not part of the active control loop, as in the case of the servo-loop described above, but exercises a process control function: in the event that the deviation between set flow rate (set value) and measured flow rate (actual value) exceeds the tolerance limits of the actuator (±4%) and the monitor (±10%), the process control system alarms the user and asks for corrective action, e.g. manual control, recalibration or repair. As a consequence of this cybernetical approach, the process monitor can be of inferior accuracy to the actuator without impairing the accuracy of the delivery system.

Factos which determine the accuracy of fresh gas metering systems are summarized in Table 3. The digital volumetric metering achieves an overall accuracy of ±4%. Overall accuracy is determined by actuator accuracy (±3%), tempera-

Table 3. Factors which determine the accuracy of fresh gas metering

	Conventional flow meter and needle valve	Analogue servo-loop metering	Digital volumetric metering
T	± 3%	< 1%	±2%
p	± 5%	< 1%	<1%
Δp	± 2%	< 1%	<1%
Gas type	±20%	< 1%	<1%
Flow	±10%	N/A	N/A
Sensor	N/A	±10%	N/A
Actuator	N/A	< 1%	±3%
Dynamic effects	± 5%	±10%	±2%
$\sqrt{\sum(\text{error})^2}$	±24%	±14%	±4%

ture correction (±2%) and dynamic effects (±2%), i.e. transients in temperature, gas supply pressure and gas conditioning; pressure and gas type effects are negligible.

The analogue servo-loop metering achieves an overall accuracy of ±14%, whereby the accuracy of the flow sensor (±10%) and dynamic effects (±10%) are the major contributors. Dynamic effects are primarily related to the electronic controller which over- or undershoots during transients. Temperature, pressure, gas type and actuator (e.g. stepping motor controlled needle valve) are negligible.

The overall accuracy of the conventional flow meter and needle valve is ±24%. Major contributors are a change in gas type and the basic inaccuracy of the flow meter. Dynamic effects from temperature and supply pressure transients, e.g. due to several users at the same gas supply source, and conditioning of the gas contribute further to the overall accuracy.

Anaesthetic Agent Metering

As in the case of gas delivery systems, three methods are known for anaesthetic agent metering: (a) conventional vaporizer; (b) analogue metering with servo loop; and (c) digitally controlled volumetric metering.

Overall accuracy of the system should be better than ×10% of the actual value; ±5% would be more desirable (see Table 4). The dynamic range covers a few orders of magnitude, considering the low flow (50 ml/min) to high flow (121 ml/min) range and the lowest (0.1 vol%) and highest (4 vol%) concentration setpoints. Data logging and remote control capability are further requirements.

Figure 4 describes a digitally controlled volumetric metering system for liquid anaesthetic agents. A fixed volume of known quantum is filled with liquid agent at a frequency f. The delivered liquid flow rate $\dot{V}_A$ is the product of volume V times frequency f.

$$\dot{V}_A = Vf \tag{3}$$

Table 4. Design objectives for quantitative anaesthetic agent metering

- Safety and reliability	Overfilling protection; agent-specific filling mechanism; safeguard against inadvertent adjustment; failure detection capability; high reliability; fail-safe design
- Overall accuracy	< ±10%
- Reproducibility	±3%
- Stability	< ±2%/month
- Response time	$T_{90} \leq 10$ s
- Dynamic range	0.1 vol% at 50 ml/min to 4.0 vol% at 12 l/min (1:10000)
- Data logging	
- Remote control	

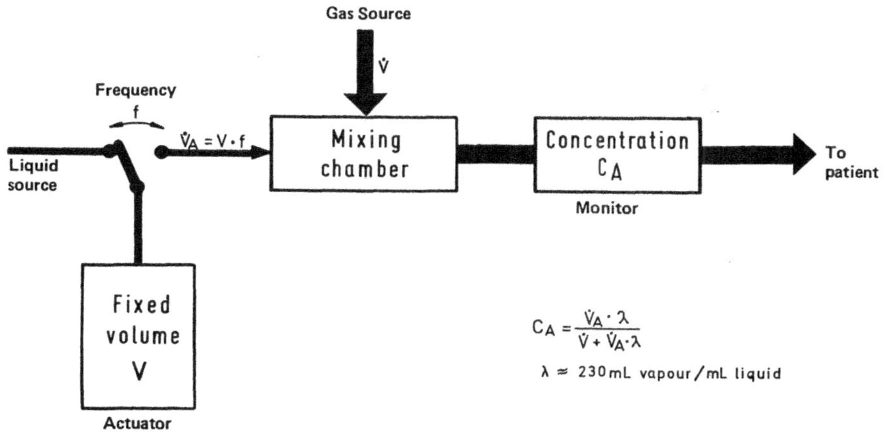

Fig. 4. Digitally controlled volumetric anaesthetic agent metering

The liquid is mixed with the gas flow $\dot{V}$ and evaporated by adding the necessary energy (heat) for evaporation.

The process is monitored by an independent concentration measurement, e.g. by an infrared analyzer, which is not part of the active control loop, as it would be in a servo-loop controlled system. It exercises a process control function.

Figure 5 shows the technical realisation of loading and unloading the fixed volume with liquid agent. The micropump operates in a four-stroke fashion:

1st stroke: loading by transfer of free moving piston
2nd stroke: both pistons and agent are transferred to dumping position
3rd stroke: dumping of agent by reversed transfer of free moving piston
4th stroke: transfer of both pistons to loading position

Table 5 summarizes the factors which determine the accuracy of the generic anaesthesia agent metering methods. The digitally controlled volumetric metering achieves an overall accuracy of ±8%, major contributors are the accuracy of the actuator and dynamic effects from transients in temperature, gas flow and evaporation and mechanical tolerances of the actuator. Temperature effects are estimated to contribute ±2%, while back pressure, gas type and steady flow effects are considered negligible.

The analogue controlled servo vaporizer consists e.g. of a bubble-through vaporizer with stepping motor driven needle valve and an infrared sensor as active part in the servo loop. Sensor accuracy of ±10% and dynamic effects of ±10% accuracy from over-and undershooting of the controller during transients result in an overall accuracy of ±14%. Temperature, back pressure, gas type and steady flow effects are considered negligible.

The conventional plenum vaporizer provides an overall accuracy of ±32%. The dependence of the vapour pressure contributes ±15% in accuracy over the

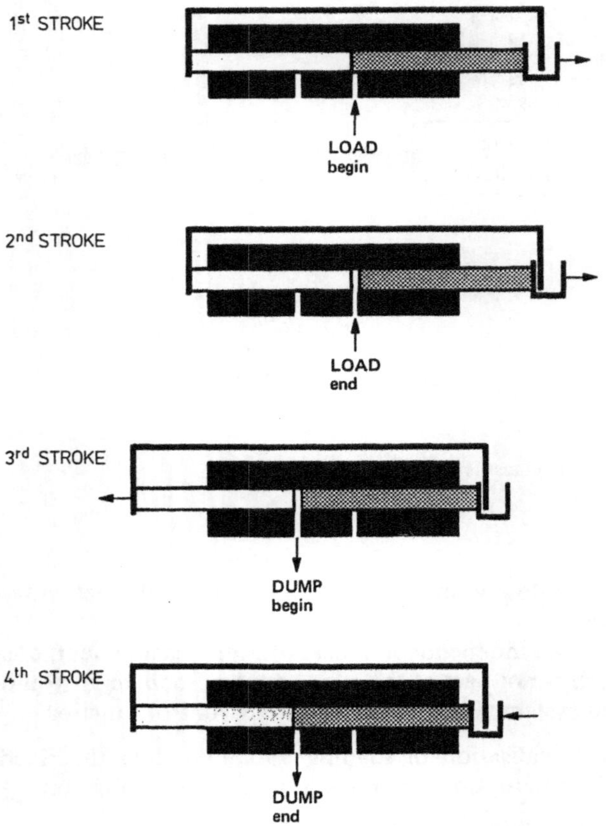

Fig. 5. Micropump for liquid anaesthetic agent

Table 5. Factors which determine the accuracy of anaesthetic agent metering

	Conventional vaporizer	Analogue controlled servo-vaporizer	Digitally controlled volumetric metering
T	±15%	< 1%	±2%
Δp	±20%	< 1%	<1%
Gas type	±10%	< 1%	<1%
Flow	±10%	< 1%	<1%
Sensor	N/A	±10%	N/A
Actuator	N/A	< 1%	±5%
Dynamic effects	±15%	±10%	±5%
$\sqrt{\sum (\text{error})^2}$	±32%	±14%	±8%

operating range, back pressure fluctuations provide a ±20% accuracy, gas type and flow contribute another ±10% each. Dynamic effects, e.g. from carrier gas changes add a further ±15%. An independent assessment by Dr. Gilly (this volume, pp. 81 ff.) using his own measurements as well as literature data for flow, temperature, back pressure and carrier gas effects yields an overall accuracy assessment of ±30% which comes very close to our own assessment of ±32%.

Volume Delivery System

Ventilators used in anaesthesia provide the patient with adequate volumes of breathing gas (see Table 6). System leaks should be detected and compensated through adjusting the delivered volume. Internal compliance compensation and fresh gas independent volume delivery ensure that the patient receives the set tidal volume. Data logging capabilities and the availability of remote control are desirable for quantification and automated recordkeeping.

For quantification digitally controlled ventilators are preferable over conventional and servo-loop ventilators, due to the improved accuracy of digitally controlled volumetric delivery of breathing gases. Such a system is shown in Fig. 6, where a defined volume is filled with breathing gas and fresh gas. The displacement of the piston is measured with a linear displacement transducer, where the displacement l and the constant cross section of the piston allow an accurate calculation of the delivered volume.

$$V_T = A\,l \tag{4}$$

This defined volume is now delivered to the patient by switching a valve from the gas source to the inspiratory line. The airway pressure measurement monitors the process.

A comparison of conventional, analogue servo loop, and digital volumetric ventilation with respect to factors determining accuracy is presented in Table 7.

Table 6. Design objectives for quantitative ventilation

– Safety and reliability	Positive and negative pressure limitation; stenosis, disconnect and reverse flow detection; leakage detection and compensation; internal compliance compensation; fresh gas independency; failure detection capability; high reliability; fail-safe design
– Overall accuracy	$< \pm 10\%$
– Reproducibility	$\pm 1\%$
– Stability	$\leq \pm 1\%$/month
– Response time	$T_{90} \leq 3$ s
– Dynamic range	$V_T = 50$ to 1400 ml (1:28) f = 6 to 60/min (1:10)
– Data logging	
– Remote control	

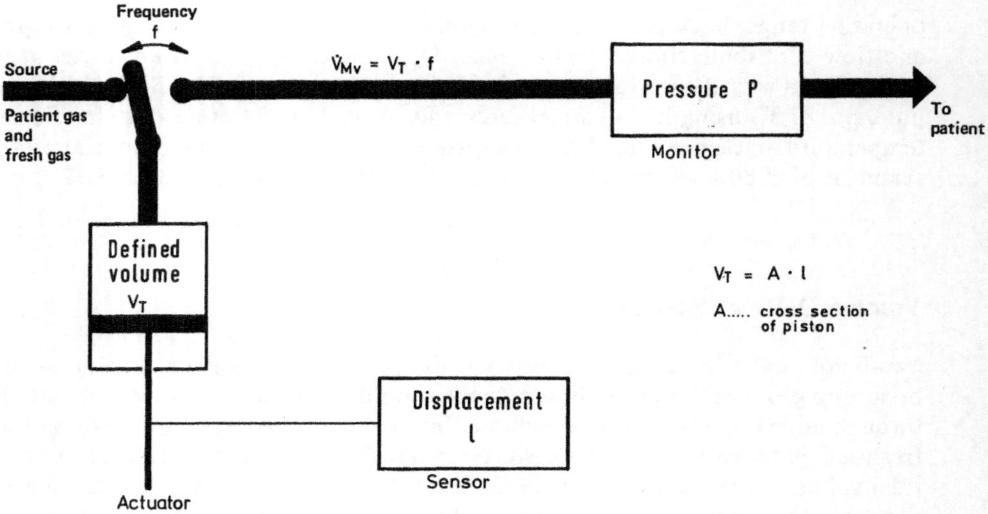

Fig. 6. Digitally controlled ventilation

Table 7. Factors which determine the accuracy of ventilation systems

	Conventional ventilation	Analogue servo-loop ventilation	Digital volumetric ventilation
T	± 5%	± 5%	±2%
Frequency	±15%	± 5%	<1%
Fresh gas flow	±15%	± 5%	±5%
Compliance	−25%	−10%	−5%
Leakage	− 5%	− 5%	−3%
Sensor	N/A	±15%	N/A
Actuator	±10%	N/A	±2%
Dynamic effects	±10%	±10%	±3%
$\sqrt{\sum(\text{error})^2}$	±29%	±21%	±7%

The most prominent factors determining accuracy in conventional ventilation systems are compliance, fresh gas flow and frequency. Dynamic effects such as non-linearities in the movement and compliant configuration of the bellows and differences between bellows also affect the overall accuracy of the system. Leakage and ambient temperature moderately influence accuracy. An overall accuracy of about ±29% is achieved with conventional anaesthesia ventilation systems. Using analogue servo loop ventilation, fresh gas flow dependence and compliance can be corrected to a certain content, however, the limiting factor here is the accuracy of the sensor required for feedback. In analogue servo-loop ventilation one encounters overshoot when changing the set point (dynamic effects). The overall accuracy improves somewhat, but ±21% is still too high for quantitative anaesthesia. Only digital volumetric ventilation provides an overall accuracy of about ±7%. This is achieved mainly through compliance and fresh gas compensation. Dynamic effects are reduced, as compared with conventional ventilation, since the compliant bellows movement is replaced through a movement of a rigid piston. Overshoot, like in servo-loop ventilation does not occur with digital volumetric ventilation.

Breathing System

The breathing system, which is the link between fresh gas source, ventilator and patient, is one of the most important parts of the system. It has to wash CO_2 out of the expired gas by an absorber, mix it with the fresh gas and deliver this mixture to the patient. To qualify for quantitative anaesthesia, a breathing system would have to meet the following requirements: low leakage; fresh gas buffering for true tidal volume; high mixing of breathing gas with fresh gas; high moisture and heat conservation; high fresh gas utilization; high absorbent utilization; high flow, low flow and closed circuit capability; small system volume and low resistance.

In quantitative anaesthesia low leakage is an essential requirement. This objective is met with compact breathing systems, where only a few fittings are necessary. The remaining features of a breathing system are primarily determined by the arrangement of its components. If one places the fresh gas inlet, e.g. near the exhaust, parts of the fresh gas will disappear unused into the scavenging system, which would impair fresh gas utilization. In addition it would make quantitative metering impossible. If the fresh gas is buffered in a bag or bellows, as shown for the first stroke in Fig. 7, fresh gas losses via the exhaust are prevented and high fresh gas utilization is achieved. Fresh gas buffering has further advantages. For example, A. Aldrete (this volume, pp. 72ff.) has shown for several American (but not limited to American) anaesthesia ventilators, that the tidal volumes delivered to the patient are perturbed by the fresh gas flow. Only for the fresh gas flow, used for calibrating the tidal volume, the set tidal volume equals the true one. Below this calibration point true tidal volume becomes smaller than the set volume and at higher fresh gas flows the true volume increases because of the extra fresh gas flow added. This effect is reduced if the fresh gas is buffered. However, to achieve true tidal volumes, the buffer volume has to be shut off from the breathing circuit during inspiration, as shown for the second stroke in Fig. 7.

Another crucial point in the design of breathing systems is the fact, that the objectives for high absorbent utilization and high moisture and heat conservation are contradictory requirements for true tidal volumes and fresh gas buffering. To meet the objective of high absorbent utilization, it is best to put the CO_2 absorber in the inspiratory branch of the system. This, however, leads to a perturbation of the tidal volume from the absorption of CO_2, leading to a reduction in the tidal volume. The absorber would also not be available as an additional fresh gas buffer. Thus, the optimization for one objective does affect one or more

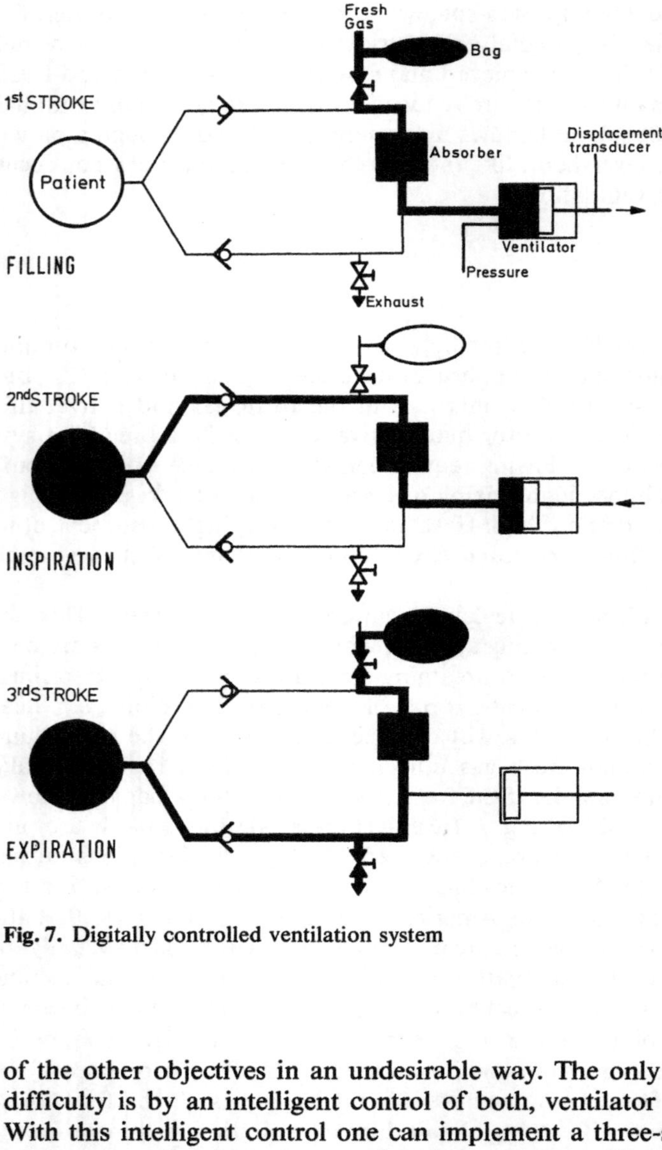

Fig. 7. Digitally controlled ventilation system

of the other objectives in an undesirable way. The only way to overcome this difficulty is by an intelligent control of both, ventilator and breathing system. With this intelligent control one can implement a three-stroke breathing cycle, while conventional systems consist of two strokes only, inspiration and expiration. During the first stroke in Fig. 7 the mixture of rebreathed gas and fresh gas is drawn from the bag through the absorber into the piston pump. During this stroke the gas is enriched with moisture and heat in the absorber for the first time. During the second stroke the tidal volume is pumped through the absorber a second time before flowing to the patient, thus resulting in an optimization of moisture and heat conservation. On the other hand, no loss in tidal volume from CO_2 absorption is suffered and the positioning of the absorber provides for additional fresh gas buffering. In the third stroke the patient expires through the absorber into the bag, so that CO_2 wash-out as well as good mixing of re-breathed gas with fresh gas are achieved simultaneously.

In summary, intelligent control of the ventilation and breathing system allows for optimization of design objectives, which would contradict themselves in conventional breathing systems. It further allows the automatic performance of very accurate leakage and complicance tests for internal corrections, which are necessary prerequisites for quantitative anaesthesia.

Summary

In Table 8 the achievable overall accuracies are summarized for the three metering systems and cybernetic methods, taking into account major environmental and controlling parameters.

Conventional systems range from ±25% to ±30% accuracy, analogue servo loops obtain ±15% to ±20%, and digital volumetric techniques provide ±4% to ±8%. The originally stated accuracy goal classifies conventional equipment as unacceptable for quantitative anaesthesia purposes. Servo-loop controlled analogue systems achieve a marginal accuracy, and only digital volumetric metering techniques come close to the desirable goal of ±5%.

However, is the application of accurate digital volumetric metering techniques alone sufficient for quantitative anaesthesia? The answer is: no. According to the definition of quantitative anaesthesia, the accurate knowledge of the quanta taken up and released by the patient is required. This can be obtained by the difference of the quanta delivered into the breathing system and leaving the breathing system. The difference is patient uptake. A problem arises from the fact that the actual amount and composition of breathing gas leaving the breathing system is not accurately quantifiable. Consequently the circuit has to be closed to eliminate the exhaust and to make quantitative anaesthesia possible.

There remains the difficulty in manually controlling the metering systems to achieve a constant inspiratory oxygen concentration, a constant end-expiratory anaesthetic concentration or a constant circuit volume by adjusting the fresh gas,

Table 8. Overall accuracy of metering and ventilation systems

	Conventional	Analogue servo-loop	Digital volumetric
Gas	±24%	±14%	±4%
Liquid	±32%	±14%	±8%
Ventilation	±29%	±21%	±7%
Summary	±25% to 30%	±15% to 20%	±4% to ±8%
Classification for quantitative anaesthesia	Unacceptable	Marginal	Desirable

anaesthetic agent and ventilation metering systems. These adjustments are tedious and very time-consuming and could distract the anaesthetist from his main responsibility, i. e. patient care.

Through remote control of metering systems, such tasks can be automated by closed-loop control techniques, as presented by Prof. Ernst (this volume, pp. 58 ff.) further discussed by Prof. Westenskow (this volume, pp. 109 ff.).

In conclusion, quantitative anaesthesia is possible with still-to-be designed digitally controlled metering systems in combination with closed-circuit anaesthesia and closed-loop control procedures.

Quintessence

The conditions for quantitative anaesthesia are:

1) Necessary condition: Accurate digital volumetric metering techniques
2) Sufficient condition: Closed circuit anaesthesia
3) Peripheral condition: Closed loop control

Quantitative Anaesthesia with the Help of Closed-Loop Control*

D. R. Westenskow and P. J. Loughlin

Introduction

Closed-loop control systems have been used since the time of the Babylonians, when irrigation systems were opened and closed in closed-loop control mode [1]. Closed-loop control in anaesthesia began in 1950 when Bickford used the EEG signal to control the infusion of thiopental [2]. A syringe with thiopental was attached to a stepper motor. Each time the EEG crossed zero, a bolus of thiopental was given. As the patient became anaesthesized the infusion rate slowed because the frequency of the EEG decreased. Closed-loop control has adjusted ventilation during anaesthesia by comparing the difference between the desired and a measured end-tidal carbon dioxide concentration [3]. If the CO_2 was high, the inspiratory airway pressure was incremented in 0.1-mm steps. When the end-tidal reached the desired level, the airway pressure was held constant. The most common application of closed-loop control is the maintenance of arterial blood pressure by the infusion of sodium nitroprusside [4]. Experience has shown that a desired end-tidal anaesthetic or blood pressure is maintained more consistently and more accurately under closed-loop control than is done manually [5]. Closed-loop systems can perform some taks of regulation better than an anaesthesiologist can. Closed-loop control can perform some very useful and helpful tasks when it is made available in the anaesthesia machine design.

When closed-loop control is added to the anaesthesia machine and breathing circuit, gas flows are controlled and provide a quantitative measure of oxygen uptake, nitrous oxide uptake, and anaesthetic uptake.

Three closed-loop controllers have been used to adjust the fresh gas flows to the anaesthesia breathing circuit [6–8]:

1) The fresh gas flow rate of oxygen to the circuit has been adjusted by closed-loop control to very accurately maintain a desired inspired oxygen concentration.
2) The volume of a closed breathing circuit has been held constant by closed-loop control of the nitrous oxide or total fresh gas flow.
3) The inhalation agent concentration, measured in the end-tidal gas, and the flow of a volatile agent have been adjusted to rapidly achieve the desired end-tidal vapor concentration.

* This work was supported in part by the ASA Patient Safety Foundation.

When closed-loop control is used to adjust fresh gas flow rates, then the circuit volume, the inspired oxygen concentration, and the end-tidal inhalation agent concentration are held extremely constant. Under these conditions (without leak or pop-off) the fresh gas flow rates are equal to the patient's gas uptake [9, 10]: the oxygen flow is a measure of the patient's oxygen consumption, the nitrous oxide flow is a measure of nitrous oxide uptake, and the inhalation agent flow is a measure of inhalation agent uptake. The use of closed-loop control and the use of a closed anaesthesia breathing system provide the conditions necessary for the quantitative measurement of gas uptake during anaesthesia.

Definition of Closed-Loop Control

In the simple control system shown in Fig. 1, the output response of a patient is adjusted according to a control input [11]. A sensor measures a feedback signal which is compared with the control input to develop an error signal. The controller uses this error signal to treat the patient in a way which reduces the error signal to zero.

Most closed-loop controllers compare the output response or state of a system with a desired or reference state. The difference between the measured and the desired state is amplified to form an error signal that tells the controller in which direction to move the system towards the desired level. Properly designed controllers reduce the error signal to zero such that the measured state and the desired state are equal.

Cosed-loop control theory describes mathematical techniques for transforming the error signal into a driving signal [12–14]. The driving signal determines the intervention, such as the rate of drug infusion, the tidal volume of a respirator or the fresh gas flow rate. The closed-loop controller chooses the correct steps and correct size of step such that the controlled variable is brought quickly to the desired level. If the steps are too large a system can oscillate, overshooting first in one direction, then in the other. To provide smooth control many closed-loop controllers use proportional, integral and derivative type control. The proportional component produces an output signal directly proportional to the error. The integral component produces a driving signal which is proportional to the integral of the error signal over time. The integral term provides a slow and steady correction to the system. The derivative component takes into account the

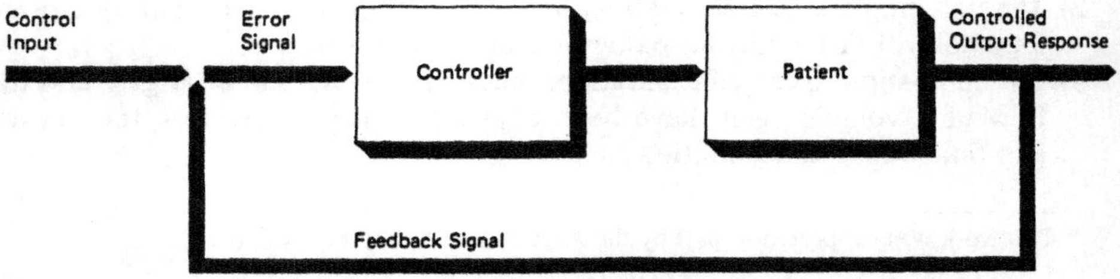

Fig. 1. Block diagram of a simple closed-loop controller. (From Westenskow [16], p. 15)

rate of change of the error signal. It provides anticipation as is needed, for example, if there is considerable delay between the drug infusion and its physiologic effect. A complete proportional-integral-derivative (PID) controller combines these three components to obtain the driving signal for the infusion device, respirator, or flowmeter.

Applications of Closed-Loop Control

Blood Pressure

Closed-loop control has been applied very successfully to the regulation of mean arterial blood pressure by the closed-loop control of sodium niroprusside infusion [4]. Several control strategies have been used, including classical PID control, adaptive PID control, and sophisticated self-tuning or adaptive control, and model reference control [15–17]. These closed-loop systems have, in general, induced hypotension more rapidly and accurately than is done by manual control. The controllers produce the desired arterial blood pressure within 1.3–6.5 min, as compared with 6.1 min for experienced anaesthesiologists [5, 15]. Pressure are maintained within 10% of the desired blood pressure 84% of the time with closed-loop control and only 67% of the time with manual control.

Neuromuscular Blockade

Another very successful application of closed-loop control is the regulation of neuromuscular blockade by the infusion of muscle relaxants [18–21]. Newer short-acting nondepolorizing muscle relaxants require close attention by the anaesthesiologist to maintain prolonged relaxation, due to their pharmacologic and metabolic properties. These short-acting drugs are best regulated under closed-loop control. Variations in patient sensitivity to the relaxants and potentiation by inhalation agents make it difficult to predict the infusion rate required to provide a given level of relaxation in a particular patient. If the level of realxation is monitored continuously, closed-loop control can be used to adjust infusion rate of the relaxant delivery according to a closed-loop control algorithm. The simplest closed-loop controller that has been employed turned the infusion rate on if relaxation was too low, and turned it off if it was too high [19]. A second controller used PID control [20] and a third used a model-based closed-loop controller [21]. Closed-loop control of neuromuscular blockade has resulted in smaller fluctuations about the desired level of paralysis and is very useful to the anaesthesiologist who uses short-acting, nondepolarizing relaxants [18].

Ventilation

Numerous systems have been developed for the control of ventilation. One system adjusts inspiratory airway pressure to control end-tidal CO_2; others have adjusted respiratory rate or tidal volume to control the end-tidal CO_2 [22–25]. Classical PID controllers seem adequate for these applications. The arterial CO_2 concentration is kept within 3 torr of the desired concentration.

Anaesthesia Delivery

Closed-loop control has been applied successfully to the control of halothane, enflurane, and isoflurane anaesthesia delivery [7, 26–30]. Inspired concentrations have been adjusted by closed-loop control to rapidly achieve and accurately maintain the desired end-tidal concentrations. Classical PID controllers have generally been used. The desired end-tidal concentrations have been reached in 1–5 min and maintained within 0.1 vol% of the desired level [8, 26, 31]. Closed-loop controlles achieve the desired end-tidal concentration much faster and much more accurately than anaesthesiologists [5].

These applications of closed-loop for control of blood pressure, neuromuscular blockade, ventilation, and anaesthesia delivery show a more rapid and precise control than is achieved by manual operators. Closed-loop control systems perform better in almost all situations.

Closed-Loop Control and Closed Breathing System

Closed-loop control can be applied with considerable advantage to the closed anaesthesia breathing circuit. It will provide better control of oxygen, circuit volume, and anaesthetic delivery while providing a quantitative measure of gas uptakes.

A closed anaesthesia breathing circuit (Fig. 2) contains a carbon dioxide absorber chamber, a breathing bag or ventilator bellows, and two one-way valves to insure unidirectional flow through the CO_2 absorber. All connections must be air-tight. Pressure pop-off relief is provided through the pressure relief valve in the ventilator.

Oxygen Concentration

An oxygen sensor is mounted in the inspiratory limb of the breathing system to measure the oxygen fraction of the inspired gas. Closed-loop control compares the measured inspired oxygen fraction with the desired oxygen fraction, the difference provides an error signal to the control algorithm [6]. The controller calculates the appropriate oxygen fresh gas flow rate to the anaesthesia system. In steady state, oxygen is added to the closed breathing system at the rate needed to keep the oxygen concentration constant. The oxygen flow provides a measure of the patient's oxygen consumption if the circuit system volume is held constant [9].

The closed-loop controller uses PID control to calculate the oxygen flow rate. The error signal to the controller is the difference between the desired and mea-

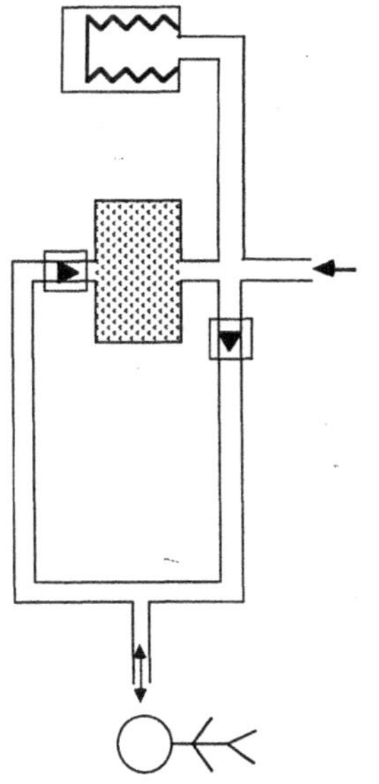

Fig. 2. Diagram of a closed breathing system

sured oxygen concentrations. The controller first calculates an output signal which is directly proportional to the error signal. It calculates a second output which is proportional to the rate of change of the error signal, and a third output which is the integral of the error signal over time. These three outputs are combined in the correct proportion to provide the controller output signal. The output signal drives the oxygen flow controller which adjusts the flow rate of fresh oxygen to the breathing system.

The oxygen controller has been shown to keep the desired inspired oxygen concentration within ±0.2 vol% of the desired value in a dog study and within 0.0±0.2 vol% in a clinical study [26, 32]. A step change is made in 1.01±0.14 min (0%–63% response) as measured in an in vitro test, 0.96±0.12 min as measured in dogs, and 1.01±0.11 min in a clinical study [32].

When steady state is reached under closed-loop control, the oxygen flow is a measure of the patient's oxygen consumption. Its accuracy is better than ±6% [9]. The oxygen control system has been used to study changes in oxygen consumption during endotoxic shock, thiopental and fentanyl infusions, dopamine, atropine, and fenylephrine infusions, cardiac pacing, deep hypothermia, and rewarming [6]. The continuous monitoring of oxygen consumption appears to help identify light anaesthesia, low cardiac output, hypoxia and decreased oxygen transport.

Circuit Volume

Closed-loop control is used to automatically control the total gas volume in a closed breathing system [6–8, 26, 31]. This control system consists of a circuit volume sensor, a closed-loop controller, and oxygen/nitrous oxide fresh gas flow regulators. The anaesthesia breathing system volume has been sensed by linear arrays of photosensitive diodes [6, 7], by LVDT transducers [26, 31] and by infrared sensors [8]. These sensors monitor the position of the ventilator bellows. They provide the signal needed to keep the circuit volume constant by measuring the bellows position at the end of expiration. By setting the desired volume at a position somewhat below the top of the bellows, the ventilator pop-off valve does not open and the breathing system remains closed. The closed-loop controller compares the signal from the circuit volume transducer with the desired circuit volume. This difference is processed through PID control to adjust the nitroux oxide infusion rate or the total fresh gas flow rate. The ratio of oxygen flow to nitrous oxide flow is calculated from the desired inspired oxygen concentration.

The closed-loop volume controller responds to change in circuit volume rapidly and immediately after a circuit disconnect. In an in vitro study, the volume was restored to the desired level in 0.47 ± 0.06 min (mean ± SD for 0%–63% response); in animals the volume was restored in 0.51 ± 0.10 min. In ten patients the response time was 0.51 ± 0.07 min. The volume remained within 0.2 ± 5.7 ml of the desired volume in the patient study. With new digitally controlled devices the volume should be restored in less than 30 s [31]. After a disconnect the circuit fills completely and automatically with the desired oxygen concentration. This feature adds a convenience to the anaesthesia machine much like automatic pilot and cruise-control.

Anaesthetic Concentration

Anaesthetic agent delivery is controlled by measuring and controlling end-tidal concentrations of halothane, enflurane, or isoflurane. Anaesthetic agent concentration has been measured by infrared absorption [7, 8, 26, 31], piezoelectric detectors [27], or mass spectrometry [29]. The patient's respired gas is sampled at the endotracheal tube connector and the concentration at the end of expiration is determined. A peak detector is used to identify the end-tidal anaesthetic concentration for each breath. This measured end-tidal concentration is compared with the desired end-tidal concentration to form an error signal which is used by a controller to adjust the setting of a vaporizer [30], the oxygen flow through a copper kettle vaporizer [26], the infusion rate of a syringe pump [7], the position of a titrating syringe [8], or the speed of an injector pump [31]. Closed-loop control adjusts the inspired anaesthetic concentration to achieve and maintain the desired end-tidal concentration. The closed-loop controller delivers anaesthesia using an automated "overpressure" technique.

Closed-loop control has been shown to induce anaesthesia in 1–5 min, with 0.3 vol% overshoot; the end-tidal concentration is held within 0.1 ± 0.01 vol% of the desired level, depending on the type of controller used [7, 25, 32, 33]. Step changes in the end-tidal desired value can be completed in 34 s [26]. When the closed-loop controllers were compared with experienced anaesthesiologists, the controller performed statistically better in 6 out of 14 categories [5]. In no case did the anaesthesiologist do better than the computer controller.

Anaesthesia Workstation

Closed-loop control is put into perspective when seen in combination with the other functions of the anaesthesia machine. Figure 3 shows how closed-loop control, monitoring, and alarms have all been integrated in a common anaesthesia workstation display. The workstation combines in one display the oxygen, volume, and anaesthetic controller inputs, a central display of monitored signals, trend displays for gas uptakes, and expert alarms. This results in an integrated system providing convenient and accurate control of closed-circuit anaesthesia. From the anaesthesia workstation, the anaesthesiologist may adjust setpoints for oxygen fractions, fresh gas flow, and the end-tidal anaesthetic concentration directly from the computer screen. He observes the monitored variables of oxygen, carbon dioxide anaesthetic agent, airway pressure, etc. During the procedure, the anaesthesiologist may call a trend plot showing oxygen consumption, nitrous oxide uptake, and anaesthetic uptake.

Our bioengineering group has developed a prototype workstation which includes Dräger devices for monitoring the concentration of oxygen (Oxycom), carbon dioxide (Capnolog), and anaesthetic agent (IRINA). Oxygen, nitrous oxide, and anaesthetic flows are adjusted by a digitally controlled metering device [34] and a saphire syringe anaesthetic micropump [34]. Closed-loop control is

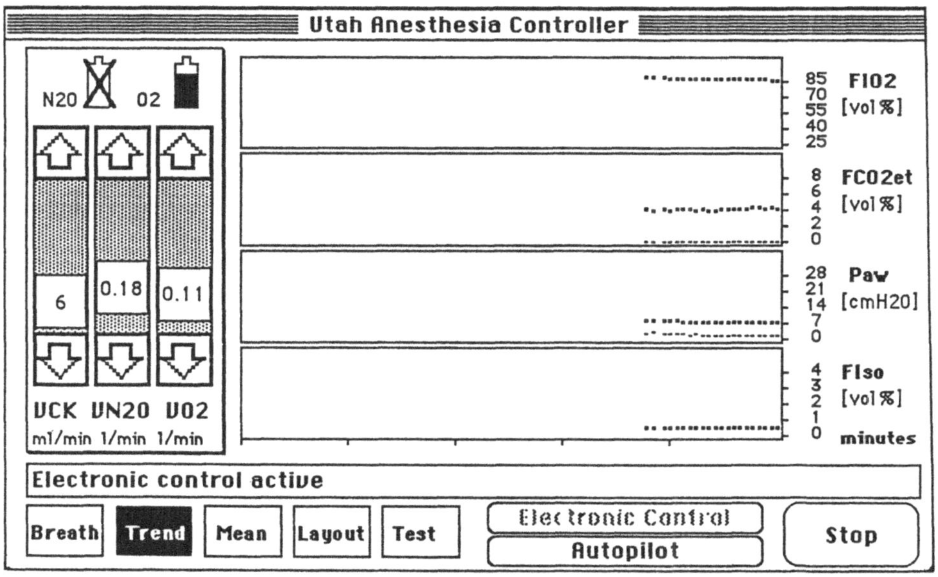

Fig. 3. Example of an anaesthesia workstation display

provided by DEC 11/23 microcomputer. The system uses a two-phase PID controller for agent delivery (halothane, enflurane, and isoflurane), a PID controller for oxygen, and a PI controller for circuit volume [31]. Using computer simulation with an uptake model [31], it was found that the controller can induce anaesthesia in 1.86 ± 1.50 min (mean ± SD of 0%–90% resonse) with an overshoot of 0.07 ± 0.041 vol%. The simulation included cardiac outputs from 2.3–6.8 l/min and ventilation from 3.6–15 l/min.

The performance of the controllers was tested using the system shown in Fig. 4. The patient breathing system was connected to the water-filled plexiglass lung model [35]. As a tidal-volume is delivered to the lung model, water moves from the sealed compartment into the compartment open to room air. During exhalation, the water returns to the sealed compartment. An air pump draws gas from the sealed compartment, bubbles it through oil, and returns it to the sealed compartment. A second pump bubbles gas through a larger oil-filled chamber and returns it to the sealed compartment (not shown in Fig. 4). These two oil chambers model the vessel-rich and vessel-poor compartments of the body, simulating a realistic anaesthetic uptake and washout [31]. Carbon dioxide is added to the FRC to simulate carbon dioxide production. Gas is removed from the FRC and replaced with 100% nitrogen or nitrous oxide to simulate oxygen consumption.

The controller was tested with the oil/lung model, inducing anaesthesia 24 times with an initial end-tidal anaesthetic concentration of 1.0 vol% halothane. The end-tidal concentration was reached in 4.4 min (SD = 1.4 min) with a maximum overshoot of 0.1 vol% (SD = 0.004 vol%). The end-tidal concentration settled to within 0.002 vol% of the desired value. Induction was rapid because anaesthesia was delivered at the highest acceptable inspired agent concentration. The controller tuned itself during the initial phase of induction so that it provided stable and accurate control of the end-tidal anaesthetic concentration throughout the procedure.

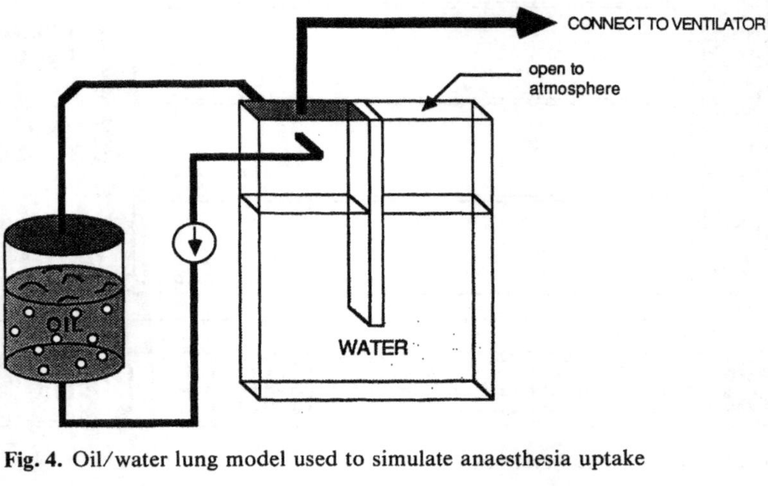

Fig. 4. Oil/water lung model used to simulate anaesthesia uptake

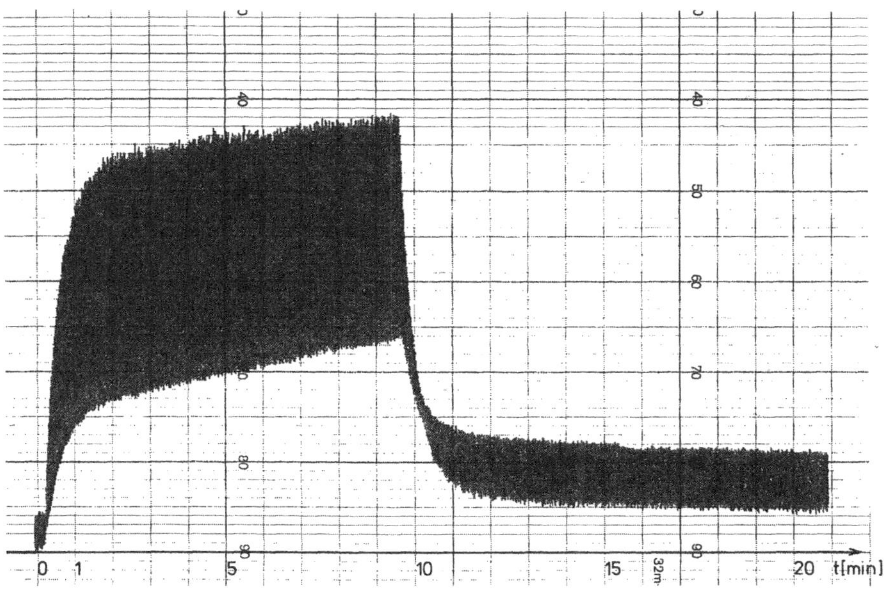

Fig. 5. Anaesthetic uptake and washout curve obtained from the oil/water lung model

The oil/lung model will find application in the further development and testing of other anaesthesia controllers. It may also find application for testing analyzer respone time [35]. For education and training, the model simulates anaesthetic uptake and gas exchange in a very realistic manner. As new devices and techniques are developed, the model may play a significant role in testing and may supplement much of the past animal experimentation.

The oil/lung model is superior to charcoal absorption because of its realistic washout curve. Figure 5 shows an example of both washin and washout. When compared with published uptake curves for halothane, enflurane, isoflurane, and N_2O, the goodness of fit is within 10%, and the shapes are much like those seen in clinical practice [10]. Although many questions about tuning the oil volumes and flow rates remain unanswered, the model will undoubtedly be a very useful tool in research, education, and device testing.

Conclusion

Only with the help of closed-loop control and digitally controlled volumetric metering devices can the anaesthesiologist so rapidly reach and so carefully maintain the desired inspired oxygen fraction, a constant volume, and the desired end-tidal concentration of anaesthetic agents in a closed breathing system.

The controllers and equipment presented in this paper provide a practical means for delivering a rapid induction, and providing careful control of the delivery of anaesthesia for quantification of anaesthetic gas, liquid, and ventilation. The workstation displays quantitative data on oxygen consumption, nitrous oxide uptake, and inhalation anaesthetic uptake which help track the patient's physiologic condition under anaesthesia.

References

1. Gadd CJ (1929) Baylonian law. In: Encyclopedia Britannica, 14th edn., vol 2. Encyclopedia Britannica, Chicago, p 863
2. Bickford RG (1950) Automatic electroencephalographic control of general anesthesia. EEG Clin Neurophysiol 2:93-96
3. Frumin MJ, Lee ASJ (1957) A physiologically oriented artificial respirator which produces N_2O-O_2 anesthesia in man. J Lab Clin Med 49:617
4. Sheppard LC (1980) Computer control of the infusion of vasoactive drugs. Ann Biomed Eng 8:431-444
5. Smith NT, Quinn ML, Flick J et al (1984) Automatic control in anesthesia: a comparison in performance between the anesthetis and the machine. Anesth Analg 63:715-722
6. Westenskow DR, Jordan WS (1984) The Utah system: computer-controlled anesthetic delivery. Contemp Anesth Pract 8:221-233
7. Ritchie RG, Ernst EA, Pate BL, Pearson JD, Sheppard LC (1987) Closed-loop control of an anesthesia delivery system: development and animal testing. IEEE Trans Biomed Eng 34(6):437-443
8. Westenskow DR, Zbinden AM, Thomson DA, Kohler B (1986) Control of end-tidal halothane cncentration. Part A: anaesthesia breathing system and feedback control of gas delivery. Br J Anaesth 58:555-562
9. Westenskow DR et al (1977) Instrumentation for measuring continuous oxygen consumption of surgical patients. IEEE Trans Biomed Eng 4(24):331-337
10. Westenskow DR, Jordan WS, Hayes JK (1983) Uptake of enflurane: a study of the variability between patients. Br J Anaesth 55:595
11. Eveleigh VW (1972) Introduction to control systems design. McGraw-Hill, New York, pp 1-8
12. Ziegler JG, Nichols NB (1942) Optimum settings for automatic controllers. Trans ASME 64:759-768
13. Isermann R (1981) Digital control systems. Springer, Berlin Heidelberg New York
14. Westenskow DR (1986) Fundamentals of feedback control applied to microcomputer instrumentation design. Int J Clin Monit Comput 3:239-244
15. Westenskow DR, Meline L, Pace NL (1987) Controlled hypotension with sodium nitroprusside: anesthesiologist versus computer. J Clin Monit 3:(2) 80-86
16. Westenskow DR (1986) Automating patient care with closed-loop control. MD Comput 3:(2) 14-20
17. Westenskow DR (1987) Closed-loop control of blood pressure, ventilation, and anesthesia delivery. Int J Clin Monit Comput 4:69-74
18. Jaklitsch RR, Westenskow DR (1987) A model-based self-adjusting two-phase controller for vecuronium-induced muscle relaxation during anesthesia. IEEE Trans Biomed Eng 34:583-594
19. De Vries JW et al (1987) Infusion of vecuronium controlled by a closed-loop system. Br J Anaes 58:1100-1103
20. Ritchie G et al (1985) A microcomputer-based controller for neuromuscular block during surgery. Ann Biomed Eng 13:3-15
21. Linkens DA et al (1982) Self-tuning control of muscle relaxation during anesthesia. In: Proceedings, IEEE conf appl adaptive and multivariable contr, Hull, England, July 1982, p 96-102
22. Frumin MJ, Lee ASJ (1957) A physiologically oriented artificial respirator which produces N_2O-O_2 anesthesia in man. J Lab Clin Med 49:617

23. Coon RL, Zuperku EJ, Kampine JP (1978) Systemic arterial blood pH servocontrol of mechanical ventilation. Anesthesiology 49:201-204
24. Ohlson KB, Westenskow DR, Jordan WS (1982) A microprocessor based feedback controller for mechanical ventilation. Ann Biomed Eng 10:3548
25. East TD, Westenskow DR, Pace NL, Nelson LD (1982) A microcomputer-based differential lung ventilation system. IEEE Trans Biomed Eng 29:736-730
26. Hayes JK, Westenskow DR, East TD, Jordan WS (1984) Computer-controlled anesthesia delivery system. Med Instrum 18:(4) 224-231
27. Ross JAS, Wloch RT, White DC et al (1983) Servo-controlled closed-circuit anaesthesia. Br J Anaesth 55:1053
28. Coles JR, Brown WA, Lampard DG (1973) Computer control of respiration and anaesthesia. Med Biol Eng 11:262
29. Mapleson WW, Chilcoat RT, Lunn JN et al (1980) Computer assistance in the control of depth of anaesthesia. Br J Anaesth 52:234p
30. Tatnall ML, Morris P, West PG (1981) Controlled anaesthesia: an approach using patient characteristics identified during uptake. Br J Anaesth 53:1019
31. Loughlin PJ, Westenskow DR (1987) A closed loop controller for end-tidal anesthetic concentration. EMBS 13-16 November, Boston
32. Furst SR (1985) A computer-controlled anesthesia delivery system. Thesis, Department of Anesthesiology, Bioengineering Division, University of Utah
33. Zbinden AM, Frei F, Westenskow DR, Thomson DA (1986) Control of end-tidal halothane concentration. Part B: verification in dogs. Br J Anaesth 58:563-571
34. Wallroth CF (1984) Technical conception for an anaesthesia system with electronic metering of gases and vapors. Acta Anaesthesiol Belg 34:279-294
35. Brunner JX, Wolff G, Cumming G et al (1985) Accurate measurement of N_2 volumes during N_3 washout required dynamic adjustment of delay time. J Appl Physiol 59:1008-1012

Quantitative Anaesthesia Management and Optimization of Machine-User Interface

N. T. Smith

Introduction

It is fitting that quantitative anaesthesia management be featured in a workshop on low-flow systems. The management of low-flow systems by the anaesthetist is not intuitive, especially with potent inhaled agents, and if performed improperly can be downright dangerous. Not only must a considerable amount of quantitative information be available to help run these systems properly, but the information must be accurate. Unfortunately, few studies have been performed on this topic, and suitable data are often not available to make useful recommendations. Thus, this essay will ask more questions than it answers.

Put simply, there are two types of machine-user interfaces – input into and output from the machine. These interfaces are focused in the automated anaesthesia record (AAR), and we shall use AAR as a paradigm in this essay.

Given the constraints of time and space, there are several topics that we shall not examine in any significant detail. These include monitors as such, both machine monitors and patient monitors. The latter are evolving so that soon they will no longer be separable from the machine. We shall also not discuss displays other than that of the AAR, again including machine displays and patient monitoring displays.

In general, we shall also avoid the topic of alarms, which is a subject into itself [1]. Perhaps a few comments will be in order, however. Historically, it has been generally agreed that alarms in the operating room (OR) – or in the intensive care unit (ICU), for that matter – have not been satisfactory.

There are several reasons for this state of affairs.

1) There are too many alarms.
2) They only have two criteria for alarming – time and/or amplitude.
3) They have often had only one state – "alarm" – rather than a set of states, e.e., "advisory," "caution," or "alert."
4) They have been totally independent of each other, that is, not integrated.

The epitome of the worst scenario occurs when several alarms go off simultaneously and ostensibly without sense. The result is that they often sound like a contemporary piece of music, something by Karl-Heinz Stockhausen, for example. The state of affairs has been so bad that a recent book devoted a significant portion of its contents to the topic of alarms [1].

There have been some preliminary, albeit noticeable, steps taken by industry towards improving this situation; this includes all of the major anaesthesia machine and ventilator companies. For example, Siemens has an infrared system for transmitting alarms to the nurse or physician, making the alarms much more private and much less obnoxious. In addition, essays by the Schreibers, father and son [2, 3], have contributed greatly to this topic.

Nevertheless, considerable improvement is still necessary. For example, integration of alarms has been implemented very little, if at all, nor has rate of change of a variable, the initial or "normal" value of the variable, or the fact that the mean or range of the variable might change as a function of what is going on during surgery or anaesthetic. As an example of the latter, blood pressure will be lower during deliberate hypotension and should be lower during aortic cannulation. Recent studies by Philip [4], Fukui [5], and Beneken [6] have attacked these problems, and their work holds great promise for vastly improved, intelligent alarms.

Proposition: There Is Too Much Information

I would like to propose that there is already too much information that is accessible *currently* both from the patient and anaesthesia system. For example, Schreiber and Schreiber [2] describe an anaesthesia machine with almost 100 controls, over 50 displays, and almost 70 alarms. And the rate of data presentation can be extremely rapid. As will be described below, somehow this information must be stored in electronic form. How much storage space will be required to store this information? Let us assume that the data must be stored at the rate of 120 Hz, with some variables an admittedly generous estimate. Each hour, then, will require 432 kbytes, or about 1/2 megabyte *per variable per hour*. This type of storage obviously requires a technology that can store large amounts of information. Most variables, of course, will not require this large amount of storage.

There has been some concern that there is too much information for the anaesthetist to assimilate. Unfortunately, this situation will become worse before it becomes better. This statement requires some explanation. First, the very drive toward quantification that is emphasized in this workshop will produce an overwhelming amount of data. There are other, perhaps more insidious and pervasive influences, however, these influences include legal pressures, hospital administrators, governmental agencies, and insurance companies. Perhaps one of the most important influences on the future surge in information is the recent establishment of standards in patient monitoring in the operating room. These standards for monitoring were established first by the Harvard Teaching Hospitals [7, 8], next by isolated component societies and insurance companies. Then by the American Society of Anaesthesiologists itself [9]. Although these standards are not laws, in our country they can have the force of laws, since any adverse result in the operating room would be difficult to defend in court if the standards had not been implemented. With the implementation of present anaesthetic standards, the number of variables and the amount of information to deal

Table 1. Possible future standards for OR monitoring

Whitcher et al. standards [9]	Frank Block standards [10]
Pulse oximeter	Pulse oximeter
Capnograph	Multigas analyzer
Spirometer	Capnograph
Halometer	Mass spectrometer
NIBP	Raman
Circuit O_2 analyzer	Circuit O_2 analyzer
Stethoscope	Spirometer
EKG	NIBP
Thermometer	Stethoscope
	EKG
NIBP (noninvasive blood pressure)	Thermometer
CINBP (continuous NIBP)	EEG
SEP (sensory-evoked potentials)	SEP
Tc (transcutaneous)	Airway pressure
	TcO_2
	$TcCO_2$
	CNIBP (Finapres)
	Cardiac output

with will become difficult to handle. Future standards could present even more of a problem, and the amount of information could become truly overwhelming. In the January, 1988 issue of the *Journal of Clinical Monitoring,* Whitcher et al. [9] provided a detailed financial argument of implementing even more tringent standards (Table 1) than are currently listed. In his accompanying editorial, Block [10] expanded this list even more (Table 1).

There are some solutions to the inevitably and rapidly increasing mass of data and information. The artificial intelligence-integrated alarms mentioned above will certainly help keep track of and sort out many of the data. They will, however, provide only part of the solution. Modeling will also provide some help in tracking and analyzing enormous amounts of data. Control systems will integrate complex information and make their own decisions regarding almost any agent that could be delivered or administered, as long as its rate of administration and its effects can be measured. These topics will be discussed below. A major solution to the mass of data and information will be AAR. And it is AAR that we shall emphasize in this essay.

How Much Information Should We Keep?

In the further development control systems, two very difficult problems will have to be worked out: the use of many variables to effect control, and the control of more than one agent at a time. At present, generally only one control variable is used, usually blood pressure. Many variables will be needed to control an agent such as dopamine, however, including arterial pressure, heart rate, cardiac output, and filling pressures. In addition, any such system of the future must be able to control many agents at once, particularly during open-heart surgery, where many agents are infused simultaneously.

There are many other problems that remain to be solved, for example, how much of the data should we store. This is not a trivial question, involving as it does philosophical, legal, and practical considerations. Traditionally, in the USA, we have stored one data point for each variable every 5 min, more frequently if ostensibly required. That interval was partly related to how difficult and time-consuming it was to acquire and record what information was available. Now, with electronic monitors, considerably more information is available and at varying rates - slowly with body temperature, relatively rapidly with an arterial pulse wave. Gravenstein et al. [11] have proposed that the recording of data every 5 min on the manual record, as suggested in the ASA standards, is not frequent enough. They have presented evidence that the frequency should be increased fivefold, to once per minute. The lawyers in our country, with their typical lack of contact with reality, would say - and have said - "Store all of it." But what is "all"? Strictly, with an ECG or arterial pressure, the original data generated by the body are analog, that is continuous, or infinite if translated into digital terms. No digital system will ever exist that can store all of the analog data. If we accept that is it impossible to save all of the data, how frequently should we save - every 5 min? - every 1 min? - every beat? Should we save just systolic, diastolic, or mean pressures, or the entire waveform? If the latter, at what frequency, or rate? Information theory suggests that two times the highest frequency component is adequate. In the case of the arterial pressure, this would be about 40 Hz.

Using the EEG as an example, another consideration arises. In order for most anaesthetists to understand the EEG, it is necessary for it to be processed, compressed, and displayed. Should one store the raw EEG or the computer-processed EEG? Less space is required to do the latter, but there can be no conversion back to the raw EEG for further processing with new algorithms, if desired. Furthermore, the choice arises as to which processing algorithm should be used. What happens if that technique or display or format or algorithm goes out of date?

How Long Should We Store the Information?

How long will the information from the anaesthetic record have to be stored? Presumably as long as the patient lives, that is, up to several decades. Thus, not only will a large amount of storage space be required for each patient, but the ultimate total amount of storage will be overwhelming, given today's technology.

How to Store Information in Patient's Hospital Records?

There are essentially two modes of storage for all this information. The first is the conventional mode - paper - with all its known problems. The second is electronic storage, with a different set of problems. It is the latter that we shall emphasize in this essay.

Electronic Memory

Fortunately some of the most remarkable changes that will soon be accessible to the everyday computer user have been in the area of memory. We shall briefly describe some of these changes here. We owe many of these advances to consumer electronics in fields such as audio and video recording and playback. Consumer electronics is driven by the possibility of enormous profits because of the mass marked involved. Therefore, considerable time and money have been spent in developing truly remarkable principles and products. In addition, because of the mass market, the products themselves are relatively inexpensive compared with other computer products. Thus, video tapes, laser technology, and optical disk technology have all contributed to the potential for storage beyond our wildest dreams. What do we do, for example, with storage that allows us to put an entire encyclopedia, or all of the textbooks in medicine onto a single disk?

One of the most exciting developments in storage capabilities is compact-disk technology, which in turn arises from laser technology. Compact-disk ROMs (read only memory) are now available for personal computers. These involve/incorporate SCSI drives, with 5-inch disks. These ROMs have 500 megabyte storage and cost $ 500–$ 1000 for the disk drive and $ 35 for each disk, for a cost of $ 0.175 per million storage bytes. Handling this large amount of stored information can be difficult for the small inexpensive computer that should be used in an anaesthetic record system. Will one be able to access all that memory in a reasonable amount of time with a small computer? The new Hypercard database system, recently released by Macintosh, can accomplish searches of millions of bytes per second. Thus, the marriage of the CD ROM technology and the Macintosh-Hypercard will advance both of these entities.

The optical/laser storage technology can be divided into several areas, according to the ability to manipulate the stored data. The least expensive and most readily available technology is ROM. In this case, the disk comes with information already on it, such as an encyclopedia or Medlines. The information cannot be altered. ROM disk drives have been selling quite extensively, and it is projected that by 1988, 250000 units a year will be sold. The next stage for CD technology is the WORM disk (write once, read many times). This disk has the advantage of allowing one to enter one's own data onto a disk, in a rather permanent, robust form. This format would be useful for storage of large amounts of data, for example, ECG or EEG waveforms. Although less than 40000 units are projected to be sold in 1988, they are now available.

Even further from full commercial implementation is the erasable CD disk, with which one can overwrite data, similar to present-day magnetic drives. However, this technology is essentially still under development, and even by 1990, only 70000 of these drives are projected to be available.

The use of the credit card-type storage format offers many exciting possibilities for storage of data. There are three types of these storage formats currently available: smart cards, laser cards, and unified cards. Smart cards contain a computer chip that provides processing power and an electronic memory that does not lose its data when the power is shut off. Most smart cards have a permanent memory of 8, 16, or 32 kilobytes; new versions use chips with reprogrammable memories. Laser cards store information as tiny black dots burned into the surface by laser in a special computer terminal. A less powerful laser that works much as one for a compact disc player reads the card. Laser cards hold as many as four million characters, much more than chip-based cards. The primary use of laser cards will be medical records. They probably will be introduced within a year.

Unified cards, as do smart cards, contain a computer chip. The unified cards have battery power, a two-line display screen, and a keyboard, so that users can enter and read information without a computer terminal. Their primary uses are financial transactions and records - again, medical records.

It might be well to reflect on the impact of 500 megabytes of memory. Five hundred megabytes is 200000 pages of text, 67 min of audio, or 5 min of video. Thus, to implement any amount of video requires the use of an interactive CD and video disk. This technology is available now.

How to Assure Temporal Compatibility?

The fact that major changes in computers, including memory storge, are taking place creates problems of its own. Thus, a major issue in the area of data storage could well become "temporal compatibility." With paper records, anyone can read the patient records of 20 years ago, assuming that the handwriting is legible. If the computer, storage medium, or software has changed, for example, this may not be possible with an electronic record. There are two choices in this dilemma: (a) Keep the same, ostensibly obsolete, computer, storage medium, or software to read the old data or (b) when any new system is implemented, transfer all data into the new system. Each of these alternatives has its own set of problems. With option 1, one could visualize accumulating over the space of many years several computer systems just to read the records from different patients. In the course of one day, all of these computers might be used. Relating the different databases could become a nightmare. And where does one enter new data on an old patient - into the original computer or new computer? Probably the former, since relating and integrating information on the same patient between two computers could be very difficult.

The second possibility, that is transfer all previous patient data to the new system, also has potential problems. Every time the data are transferred, whatever the new component of the new system, the transfer will be long, tedious, expensive, and very possibly prone to error. There could well be an incompatibility of storage or database format, given that there may be a new storage medium, a new language, or a new version of an old language.

Vital Importance of Correct Information

It is difficult to overestimate the importance of accurate information in the context of this workshop. For example, the very reasons that make the anaesthesia record important also make accuracy important. Accurate information may be even more important for control and for modeling, however. The anaesthetist can edit the information that has been entered into the anaesthesia record. A control system, on the other hand, cannot. Unless exquisite algorithms for detecting artifact and other types of erroneous information have been implemented, the controller will use any incorrect information to effect incorrect control. Similarly, if a model is used to aid servo control, or to predict variables that are otherwise difficult or dangerous to measure, incorrect information could be hazardous.

Type of Accuracy

Parenthetically, there are essentially two criteria for accuracy: that the information be artifact-free and that it be correct. Unfortunately, information that is artifact-free may not necessarily be correct, and this obvious fact is often forgotten by both.

How to Deal with Artifacts?

How does one deal with artifacts if the AAR is not an integrated system, such as that made by anaesthesia-machine companies? In this case, the receiver (AAR) depends on the transmitter (monitor); if the monitor incorrectly identifies a signal as artifact or nonartifact, the receiver may do so also. This is in addition to the difficulties in interfacing described elsewhere. If the raw signal itself is transmitted to the receiver, the latter could conceivably make its own decision on the presence of artifact, but what should be done if the two entities disagree? If the decision is made that the signal is artifact, what should the AAR do – label it as such, or omit the value?

Will AAR Solve Some Problems?

The AAR has an excellent chance of solving many of the problems of information overload.

Automated Anaesthesia Records

Uses for AAR

Before we can go into the uses and advantages of the AAR, we must first describe the uses and advantages of the anaesthetic record. The anaesthesia record has a wide variety of functions. Understanding these functions is necessary for understanding the need for effective means of entering information.

These uses include the following.

1) A Preanaesthetic Summary. Considerable demographic, physical, physiological, pathological, clinical, laboratory, and radiological information is summarized on the anaesthetic record, usually on the reverse side. Other miscellaneous information, including preanaesthetic medication and its impact, the patient's preoperative medicines, and direct medicolegal considerations, such as site of surgery and signing of consent form, are also recorded.
2) A Reminder of Information to Record. Many anaesthetic record forms are crammed with blank boxes to check, as well as blank spaces to fill out with specific information. If the anaesthetist is reasonably conscientious, this type of record can serve a very useful purpose.
3) Trends. If the anaesthetic record is kept current during the case as events occur, the anaesthetist can detect trends in blood pressure or amounts of drug, for example. From these trends he can make specific diagnoses or decide upon therapeutic maneuvers.
4) Prediction. As a corollary of no. 3, one can predict, with variable accuracy, the intra- and postanaesthetic curse of the patient. Unless the trends are clearly presented by the record, however, prediction can be difficult.
5) An Aid in Postanaesthetic Care. Frequently, diagnoses and decisions made in the recovery room or the intensive care unit are based on events occurring or agents administered during the anaesthetic. For example, in assessing postoperative apnea, it is useful to know whether and how much narcotic and neuromuscular-blocking agents were administered.
6) An Aid to the Next Anaesthetist. If the patient must receive another anaesthetic, whether days or years later, the anaesthetic record becomes extremely valuable to the subsequent anaesthetist. Perhaps of all users of the anaesthetic record, the subsequent anaesthetist is the most grateful for a complete and accurate record.
7) Teaching. The anaesthetic record is a useful teaching device, whether during anaesthesia, immediately postoperatively, or as a demonstration or a source of information at a case-presentation conference.
8) A Database. The anaesthetic record can help implement databases. The potential information available from such databases is vast, including the number or types of cases done during any period, the number of cases done by each anaesthetist, the anaesthetic agents and techniques used, and the incidence and types of complications.
9) Research. If properly constructed, the anaesthetic database can be a valuable research source. This often requires a record with more detailed requests

for information than usual. In addition, the anaesthetist must meticulously fill out every requested detail.

10) Medicolegal. In the United States, particularly California, the anaesthetic record is an important medicolegal document. Lawyers generally agree that defendant anaesthetists lose more malpractice cases on the basis of an inadequate or incomplete anaesthetic record than from almost all other causes combined.
11) Financial. The anaesthetic record contains the information necessary for billing the patient, not only for the anaesthetic charges but also for many of the hospital charges.

These many uses for the anaesthetic record have determined the type and quantity of information recorded, as well as its repetition rate and its format for display.

Advantages of AAR

In addition, if one takes into consideration these uses for the anaesthetic record, it becomes clear why an automated anaesthetic record could be very useful. An automated record keeping system provides the benefit of:

1) more accurate data recording [12];
2) less need for manual charting during crises or other busy periods of anaesthetic care;
3) a centralized display on which all data, current and trended, are available on one common screen or other display; and
4) a legible printed record at the end of the case.

Because of the obligatory detail and accuracy alone, automation of the anaesthetic record could be of great benefit to the anaesthetist. Zollinger et al. [12] demonstrated that anaesthetists' charts are less complete and accurate when compared with the real data generated. Certainly an automated record could be neater and more legible than most anaesthetic records generated manually.

Because of its neatness and accuracy, the automated anaesthetic record will be useful medicolegally. Many anaesthetists are concerned, however, that by being objective and more nearly complete, the automated record might reveal any mistakes - real or imagined. Actually, close to one-half of malpractice cases are lost not because the anaesthetist did anything wrong, but because his record was so incomplete that the plantiff's attorney was able to cast doubt on the abilities or veracity of the anaesthetist. Thus, many defense attorneys feel that an accurate automated record will win many more cases than it will lose. If the automated anaesthetic record does indicate that an error of commission of omission has occurred, the case could be settled out of court, saving time, money, and harassment of everyone - except the lawyers.

Currently, abut 10%–15% of the anaesthetist's time is spent on the anaesthetic record in the longer cases, perhaps even more in shorter cases. Given the increased number of variables that will be recorded and amount of data that will be generated, it is very probable that the relative time spend on the record will increase.

Possible Inputs into AAR

All the information that will be processed and displayed by an AAR must somehow be entered into a computer. That obvious fact can create problems. To further complicate matters, another obvious fact is that if we are to have a fully automated record, all inputs into the computer should themselves be automated. This ideal situation is not easy to achieve.

Under optimal circumstances, these inputs would include the following.

1) Physiologic Variables: During both invasive and noninvasive monitoring, when many physiologic variables are being generated electronically, the automatic transfer of physiologic information to the anaesthetic record is relatively straightforward. If the monitoring is not electronic, and particularly if it involves the senses of hearing, smell, or touch, transfer becomes more difficult.
2) Drugs: Drugs represent a more difficult situation, and many methods for transferring drug information have been proposed. One of these, described by the group at Duke University is a bar-code method [13]. This system does effect electronic transfer. On the other hand, the bar-code system used at that time was difficult to learn, cumbersome to use, and unreliable [13]. A new user-friendly and reliable bar-code system, recently displayed at the ASA meeting, claims to provide the accuracy and simplicity that has been previously lacking.
3) Infusions: Infusions are somewhat easier to deal with in an automated record system than are bolus injections, since there are available infusion devices with an electronic output that indicates the rate of infusion. Obviously, information about the drug, including its concentration, must be entered so that the infusion rate can be converted into a drug administration rate, as well as into a cumulative amount. Other information interchange is possible, and our interface with an infusion pump gives such messages as "I'm working," "bag empty," or "air lock." In addition, it will ask if the wrong concentration is being used.
4) Anaesthesia Machines: Relatively soon microprocessor-controlled anaesthesia machines will be available. It will then be a simple matter to transfer information – such as gas flows, vapor concentration, disconnects, or the amount of gas remaining in the sources – from the anaesthesia machine to the automated record. In the meantime, it will be necessary to transfer this information manually.
5) Ventilators: Similarly, ventilators are evolving into microprocessor-controlled devices, which will be able to transfer information, such as rate, I:E ratio, resistance, compliance, or wave form from pressure, or flow to the automated record. As with anaesthesia machines, until the new ventilators are available,

the information must be entered manually, assuming it is even available to the anaesthetist.

6) Annotations: There is a considerable amount of miscellaneous material that must be entered into the anaesthetic record. This material includes the pre-anaesthetic information described above, as well as description of anaesthetic techniques, such as a spinal or the insertion of an endotracheal tube; the position of the patient; and the onset of anaesthesia or surgery, or the inflation of a tourniquet. Currently, this information is usually entered manually, either in writing or by keyboard. A local area network connected with a hospital information system could automatically supply much of the preanaesthetic information from such diverse locations as admissions, the clinical laboratory, the radiology department, and cardiology. Other possible alternatives for entering this information are described below.

Monitoring is currently the most important source of quantitative information, albeit this information is only indirect as regards the machine-user interface. I have previously stated that currently the AAR represents the best mode to integrate and display this information. There are problems, however, that can make the interface between monitors and AAR difficult. These problems center on the fact that each manufacturer of monitors seems to have its own interfaces, including its own connectors, output voltages, handshake protocols, etc.

Different Modes of Entry into AAR

As discussed above, much of the information that is entered into the anaesthetic record cannot be automatically entered in the foreseeable future. Currently, this information comprises everything except physiologic measurements, and often even the latter must be manually entered. The problem for the automated record is to enter the information into the computer that is implementing this automated record – and that will ultimately aid in the storage of the data.

At least six means of information entry have been proposed:

1. The conventional keyboard;
2. A special-purpose keyboard with built-in function keys for single key-stroke entry for an entire drug name, for example;
3. A special pad on which to manually print letters and numbers, with the printed characters being entered into the computer;
4. A bar-code mechanism;
5. A touch-screen;
6. Speech recognition by a computer.

We shall briefly examine each of these techniques. The conventional ("QWERTY") keyboard would seem to be the most appropriate and useful mode of entry, since an infinite variety of information can be entered with it. Yet there has been a genuine resistance to its use by clinical anaesthetists. The absolute flexibility of the typewriter keyboard is also its disadvantage. Typing is a time-consuming diversion of the clinician's attention from the patient. Certainly its facile use almost requires the use of two hands; most of us can type several times faster with two hands than with one hand. Such distraction of hands or mind can be disruptive, particularly when the patient's condition is unstable, which is, unfortunately, the same time that man-machine communication may be most needed. The use of standard keyboards in experimental computerized monitoring systems in the past has been a major stumbling block to the acceptance of these systems. Nevertheless, the increasingly general use and availability of computers – and their keyboards – may bode well for the eventual acceptance of the conventional keyboard. Another encouraging development is that of a new keyboard layout (Dvorak) which permits much more natural and, hence, rapid typing. Although those accustomed to the QWERTY keyboard may find the new one difficult to adjust to, beginners, or those who have assiduously avoided typing, may find it the stimulus needed to use a full keyboard in everyday, as well as in operating-room life. On the other hand, perhaps the typewriter keyboard's use will be reserved for more complicated messages entered at the anaesthetist's leisure.

A variation that has been used in several experimental systems with different degrees of success is the special-purpose keyboard. This keyboard usually has fewer keys by far than the standard QWERTY board, and each key represents an entire entity, such as the name of a drug, or a function. The concept is to decrease the amount of typing required to send the message to the machine. The special-purpose keyboard would seem to be ideal, since one can enter an entire word or phrase with one key stroke. However, to incorporate onto one keyboard all the words and phrases necessary or useful to the anaesthetist (about 350) is an impossible task. Therefore, one must either severely limit the number of available words, or design for sequential entry. With the latter, one strikes a special key for a class of drugs, for example. Thus, one could enter "barbiturates" and obtain a list of barbiturates on the screen. Next, one would enter a number for the appropriate agent, then enter the route and dose. This requires access to and attention to the computer and/or screen – not always a convenient matter. The remote key pad is a possible solution to this dilemma.

A means to directly enter hand-printed information into the computer would seem to be ideal, since one does not need a complex keyboard, and since one can use a very familiar mode of entry – printing. More importantly, perhaps, one can enter information onto a form very easily, since the entry pad may incorporate a form, and the location of an entry on the pad determines its location on the computer "form" on the screen. However, the system is relatively cumbersome to set up. Furthermore, printing is a laborious task for many, and requires close attention to the special pad and to the screen.

Perhaps most importantly, most current systems can recognize only the neatest of printing, thus making them difficult for this author, at least. Neural networks are good at pattern recognition, and newer pads incorporating this technology should at least solve the last problem. These systems are easily trained, and perhaps they can be trained to recognize individual handwriting, rather than printing, which is slower.

If we agree that one of the main goals of the automated record is to free the hands of the anaesthetist to accomplish other tasks, then none of these modes of entry is ultimately satisfactory. Furthermore, the hands of the anaesthetist are often occupied during the time that information is being generated at the most rapid rate, for example, during induction of anaesthesia. We, therefore, have been evaluating the possibility of using voice-recognition technology to help implement an automated record keeping system. With a voice-recognition system, the anaesthetist can enter information, such as a blood pressure value or the insertion of an endotracheal tube, while taking the measurement or performing the act, without the necessity of interrupting the routine or act to enter the information. Voice-recognition technology has reached the stage of development that allows us to examine the feasibility of its use for the automated record.

Voice Recognition

The technology that allows this hands-off entry into the anaesthetic record is called computerized speech recognition (CSR). An initial overview of the operation and constraints of CSR will help the reader understand the subsequent detail on the CSR.

Voice recognition most commonly makes use of a microphone, specialized software, and a plug-in expansion board containing a microprocessor for speech-to-computer data conversion. This is used in conjunction with a "host" microcomputer, that acts as a controller.

Use of most voice-recognition systems begins with a 10- to 15-min training session, during which the user speaks various key words into the microphone as they appear on the computer screen. Thus, the system acclimates to the accent, articulation, and timbre of its "master's" voice.

In all speech recognition systems the user (speaker) chooses his words from a predefined vocabulary of usually up to 100 in a set. The computer's task is to identify, or recognize, which of the vocabulary words are being spoken. During a brief pause after each word the computer recognizes that word. Most systems are speaker-dependent, limiting the system to one speaker at a time and allowing an individual user to tailor the system to his voice by "training" the computer – i.e., pronouncing the entire vocabulary for the computer – before first use.

For computerized voice input, processing chops up the pieces of sound in a spoken word and analyzes and records the waveform of audio frequencies in each piece. One method of analyzing the pieces of sound uses filters that sample sound at specific frequencies (much like the graphic equalizer in a car stereo) and record each level.

The SRC operates in two modes: training and recognition. During either mode, the SRC processes the sine waves of each word of speech from the microphone into a linear stream of zeroes and ones, digital binary pattern, of uniform length. This pattern is called a token. During training, the tokens are used to make templates, each corresponding to one vocabulary word, and each stored in memory as a reference pattern. Algorithms are programmed into the SRC for generating the tokens and the templates and for matching a token against the templates. In addition, there is a communication protocol by which the SRC interfaces with its host system. These algorithms differ for each commercial device, and the details are usually proprietary. Nonetheless, the following general description applies to several SRCs, including that used in the EARS system. The microphone signal, after suitable preamplification and broad filtering, is sent to a bank of band pass filters that collectively span the frequency spectrum of speech. The outputs of these filters are digitally sampled and combined into composite spectral snapshots. Those that differ little from their predecessors are discarded so that weight is given to periods of spectral change, a process known as time-warping. Finally, the surviving snapshots are combined into a single pattern that is shrunk or stretched to form a token of a standard, predetermined size. Thus, the token represents a synopsis of the time-varying spectral analysis of a spoken word. To convert speech into tokens requires considerable data reduction to make computer analysis feasible. Yet these clever algorithms manage to retain enough information to capture the character of each word.

We have developed a CSR system called EARS (Entry into the Anaesthetic Record by Speech) [14, 15]. The remainder of this section describes this system, its operation, and its benefits and drawbacks, as well as those of speech recognition in general in the operating room.

The Speech-Recognition Computer

During recognition in our system, each utterance produces the token that is compared with each template stored in memory. Each comparison yields a score indicating the degree of similarity, theoretically ranging from perfect match to no match, although either extreme is unusual. The computer selects the template with the highest score. In the unlikely event that two or more templates tie for highest score, one is arbitrarily selected. Because there is always a best score, no matter how poor, the highest score can be compared with a threshold value, to determine acceptance or rejection of the word. This helps prevent an inappropriate or ambiguous utterance from being incorrectly recognized. The second-best score identifies the runner-up word. The difference (delta) between the winner and runner-up scores can serve both as a rejection criterion and as a measure of conficence. A high delta indicates that the winner had a much higher score than any other word and hence is likely to be correct, whereas a low delta indicates that the winner and runner-up were nearly identical, casting some doubt on the winner's correctness. Making the rejection criteria too loose favors misrecognition of unintentional or sloppy speech or of ambient operating room noise; making them too tight favors rejection of valid utterances. Optimizing this trade-off is a matter of trial and error.

Applying Syntax

During recognition the fewer the words in a vocabulary set, the fewer templates with which a token must be compared and the greater the likelihood of correct word identification. Certainly it is easier to find the best match from among, for example, a dozen templates than from among 100. By means of a command from the host computer to the SRC, recognition can be restricted to a given vocabulary subset by comparing a token with only the templates of specified words. A fundamental strategy of programming the host computer, therefore, is to restrict recognition to as few words as possible at one time. To accomplish this strategy we set up a special "Language for Anaesthesia Record-Keeping" (LARK). This language consists of a vocabulary and a syntax. In any language a single word rarely conveys a message. Meaningful speech consists of strings of words organized into sentences. Since words are not chosen at random, rules must exist to specify how words may be combined into sentences – i.e., which words may follow which. The collective set of rules is called the *grammar,* or *syntax*, of the language. The syntax of a natural language such as English is immensely complex and subject to continual debate and revision. For LARK, the situation is much simpler, since we can define and maintain a "private" language. To set up LARK, we first divided the large anaesthesia word group into several smaller but still major word groups such as monitors, drugs, events, and fluids. The EARS program initially instructs the SRC to listen for one of these major words. After one of them is spoken and recognized, the computer will call up the subset of words related to that key word. After a word is recognized from this new subset, still another subset is specified, and so on. Recognition proceeds from subset to subset, always anticipating candidate words, until a syntactically complete and correct sentence is built. The calling of a subset is a rapid process. Sentences are terminated by a long pause if additional words are permissable, or automatically if no additional words are permissable.

An example of a sequence through the syntax follows. Enunciating the key word "measurement" will bring up into the recognition memory words such as "blood pressure" or "heart rate." After one of these words is spoken, the "numbers" group of words will be called up, succeeded if appropriate, by the units group (and the administration route group for drug groups). To permit entry of arbitrarily long numbers in a system with vocabulary constraints, we use single digits only; the number 110, for example, is announced as "one, one, zero." Thus, a typical LARK sentence would be "Measurement blood-pressure one one five over seven five." Other sample sentences include: "Event tourniquet down," "Airway oral endotracheal tube seven point five," and "Breathing nitrous-oxide three liters per-minute." The hyphenated pairs of words in the preceding sentences are each spoken and recognized as single words. EARS has a vocabulary of about 350 words, with about 120 syntax rules. The syntax is designed to achieve some flexibility. For example, "Monitor precordial stethoscope" and "Monitor stethoscope precordial" are equally valid sentences. "Drug morphine four" and "Drug morphine four milligrams IV" are equally valid and have the same meaning, because omitting the unit and route implies defaults of milligrams and intravenous, respectively, simplifying most drug entries.

System Performance

For CSR to be viable in real-world applications, problems such as noise rejection, variation in pronunciation, and error correction must be dealt with. In addition, there must be ways to accurately and meaningfully measure performance. How does the SRC deal with the enormous diversity of human speech, particularly in the specialty of anaesthesia, which is a specialty of dialects? To begin with, it is another advantage of speaker-dependent systems to accomodate any dialect or accent, since the templates are unique to the individual user. Indeed the entire vocabulary could be translated, word for word, into a different language, and the SRC would be indifferent.

Speaker-dependent systems have other significant advantages, compared with speaker-independent ones, aside from their greater vocabulary and accuracy. Someone else's voice saying the right word at the wrong time will not upset speaker-dependent systems. Another advantage of speaker-dependent systems is the capability of speaker verification. With this technology, the voice functions as a fingerprint to ensure that only an authorized person has access to a patient's record, and, certainly, that only an authorized person can change it. In addition, it would provide a verification as to who had entered into or altered a record. Nor is a limited vocabulary a disadvantage. The fewer words available for recognition, the fewer possibilities of extraneous words deceiving the system.

Other aids to better recognition include automatic gain control at the microphone input, which can compensate for variation in overall loudness. The microphone is an important component of any system, and microphones that "hear" almost maintain the actual record. It is best regarded as an alternative to a keyboard, one that convers an unusual degree of physical freedom.

Future of Voice Recognition

Currently, and for the near future with voice recognition, the major consideration is evolutionary, rather than breakthrough, technology. An important advance is the availability of systems that may achieve 99.98% accuracy even under high levels of environmental noise, such as in the cockpit of a helicopter at 100 dB. These new systems can also process connected speech, as well as perform word spotting, which detects vocabulary words out of a stream of random input. Parallel processing is also beginning to have an impact. For example, individual special-purpose processors on the voice processor can perform parallel tasks of digital signal processing and pattern matching. Processing power has increased dramatically. The sophistication of recognition algorithms has increased. For example, changes in voice caused by stress are tolerated better. Finally, the cost has dropped dramatically. These advancements are already apparent. A typical voice-recognition system built in 1982 with a 100-word vocabulary sold for about $ 2000, while units with ten times the word capacity retailed for nearly $ 10000. Today you can have the same capability for less than $ 300. Soon, we can look forward to faster, more reliable speech-recognition boards in the $ 100 to $ 200 price range.

The more distant future is even more optimistic. A number of companies are currently engaged in serious competition to develop a voice-activated word processor, a speaker-independent, continuous-speech recognition system. The technology that may well be used in this processor may well be neural networks. The "voice writer" will translate spoken input directly into neatly typed text. Its vocabulary will be in the tens of thousands of words, and it will have sufficient computing power and speed to recognize a variety of speech patterns and accents. These corporate projects are aiming at a completion time of less than 5 years. The "voice writer" will have a significant impact, when it arrives. And neat, accurate, detailed records will no longer require fast touch-typists.

Control

This workshop has emphasized servo (closed-loop) control, and for good reason. The same reasons that make the availability of quantitative information so important also apply to closed-loop control: low-flow anaesthesia is not intuitive and can be dangerous if not performed properly. Closed-loop systems have the advantage that they can watch the patient very closely, and even anticipate problems before they occur.

Nevertheless, the implementation of closed-loop control during low-flow anaesthesia can be difficult, since there is a long time interval (*delay*) between a change in anaesthetic concentration and the effect of that change. Imagine trying to drive an automobile where the wheels turned 30 s after one turns the steering wheel!

Pioneers in the area of servo control for low-flow systems were Coles and Lampard in Australia [16–19]. A preliminary report describes a ventilator that switched on pressurized anaesthetic gases for inspiration and opened a low impedance path to atmosphere for expiration. A unique feature of the system was the use of central venous pressure to monitor an overdistention of the lungs, as determined by impaired venous return. The ventilator was used to maintain a constant end-tidal pCO_2 in sheep. Subsequently, the Australian group developed techniques for controlling the concentration of O_2, the volume in a closed-circuit system [16], and the infusion of a neuromuscular blocking agent via an evoked electromyogram.

The difference between changes in inspired vapor concentration and actual anaesthetic effect can be very large with closed-circuit anaesthesia. We, therefore, decided to investigate an effect of the anaesthetic itself. It also seemed to us that it was usually not easy to predict without a model the effect of changes in inspired anaesthetic vapor concentration on anaesthetic depth, given the many possible variations among patients. It seemed that the best approach to the problem would be to use the EEG as an indicator of anaesthetic depth. In essence, we wanted to see if the EEG could be used by itself to aid in the adjustment of inspired anaesthetic concentration to modulate anaesthetic depth during closed-circuit anaesthesia with halothane.

The protocol was relatively simple. We studied six patients undergoing dental procedures. We monitored two leads of the EEG with the Density-Modulated Spectral Array (DSA) [20, 21] implemented with our Biomedical Computer (BMC) [22, 23], as well as the ECG, cuff blood pressure, heart rate, and gases with a Perkin-Elmer mass spectrometer. An anaesthesia resident saw only the DSA and an out-of-circle vaporizer. He was not even allowed to watch the surgeon or the patient. Given his limited information - one input and one output - he was supposed to control the depth of anaesthesia. The study did not begin until the patient was well anaesthetized, and the resident was supposed to maintain anaesthetic depth as constant as possible. He was trained by showing him two "boundary" EEG patterns; he was to maintain the EEG between those two patterns. The staff had access to all of the information listed above, including the patient and the surgery. He also had the power of veto, in case anything went wrong.

We selected dental cases so that there would be little in the way of surgical stimulation or blood loss, thus making the assessment of anaesthetic depth more valid. We judged "true" anaesthetic depth via three physiological variables: arterial pressure, heart rate, and end-tidal pCO_2. Table 2 demonstrates that the resident was able to maintain these variables - and therefore presumably anaesthetic depth - reasonably constant with the EEG alone over a period of almost 2 1/2 h. In addition once the surgery was completed, the patient awakened in less than 5 min.

Models

Just as with quantitative information and with servo control, the same considerations apply to the importance of models in low-flow anaesthesia: And once more, the importance of correct, accurate, and complete models cannot be overemphasized. Models usually represent an average patient. On the other hand, it is possible to adapt an "average" model to the patient being anaesthetized or treated. If this is done, many new doors are opened.

Summary

This essay has explored some of the considerations involved with the anaesthesia machine-user interface. We have concentrated on data, especially the implications of the enormous amount of data that will be interchanged in the not-

Table 2. Closed circle with EEG controlled ventilation with halothane/O_2 ($N=6$)

Time from induction (min)	39	71	141
SAP	99 ± 8	104 ± 9	107 ± 9
HR	75 ± 18	79 ± 17	81 ± 16
$PACO_2$	32 ± 7	37 ± 6	33 ± 6

too-distant future. The automated anaesthetic record was discussed in detail as being useful for solving some of the problems, present and future. Voice recognition in particular can help input into the computer data that otherwise must be entered manually.

References

1. Gravenstein JS, Newbower RS, Ream AK, Smith NTy (eds) (1987) The automated anesthesia record and alarm systems. Butterworths, Boston
2. Schreiber P, Schreiber J (1987) Anesthesia systems – risk analysis and risk reduction. North American Dräger
3. Schreiber P (1985) Safety guidelines for anesthesia systems. Merchants, Boston
4. Philip JH (1987) Intelligent detection of artifact. In: Gravenstein JS, Newbower RS, Ream AK, Smith NTy (eds) (1987) The automated anesthesia record and alarm systems. Butterworth, Boston, pp 175–190
5. Fukui Y (1987) An expert alarm system. In: Gravenstein JS, Newbower RS, Ream AK, Smith NTy (eds) (1987) The automated anesthesia record and alarm systems. Butterworths, Boston, pp 201–210
6. Beneken JEW, Gravenstein JS (1987) Sophisticated alarms in patient monitoring: a methodology based on systems engineering concepts. In: Gravenstein JS, Newbower RS, Ream AK, Smith NTy (eds) (1987) The automated anesthesia record and alarm systems. Butterworths, Boston, pp 211–228
7. Eichhorn JH et al (1986) Standards for patient monitoring during anesthesia at Harvard Medical School. JAMA 256 (8):1017–1020
8. Hornbein TF (1986) The setting of standards of care (editorial). JAMA 256 (8):1040–1041
9. Whitcher C et al (1988) Anesthetic mishaps and the cost of monitoring: a proposed standard for monitoring equipment. J Clin Monit 4 (1):5–15
10. Block FE (1988) A proposed standard for monitoring equipment: What equipment should be included? (editorial). J Clin Monit 4 (1):1–4
11. Gravenstein JS, deVries A, Beneken JEW (in press) (1989) Sampling intervals during anesthesia. J Clin Monit 5 (1):14–18
12. Zollinger RM, Kreul JF, Schneider AJ (1977) Man-made versus computer-generated anesthesia records. J Surg Res 22:419–424
13. Block FE et al (1985) Two computer-based anesthetic monitors: the Duke automatic monitoring equipment (DAME) system and the microdame. J Clin Monit 1 (1):30–51
14. Smith NTy (1985) Automated anesthesia record keeping. In: Osswald PM (ed) Computers in critical care and pulmonary medicine. Springer, Berlin Heidelberg New York, pp 147–161
15. Smith NTy, Quinn ML, Sarnat AJ (1987) Speech recognition for the automated anesthesia record. In: Gravenstein JS, Newbower RS, Ream AK, Smith NTy (eds) (1987) The automated anesthesia record and alarm systems. Butterworths, Boston, pp 115–134
16. Coles JR, Brown WA, Lampard DG (1971) Computer control of respiration and anesthesia. Digest of the 9th international conference on medical and biological engineering, p 141
17. Lampard DG, Coles JR, Brown WA (1973) Electronic digital computer control of ventilation and anaesthesia. Anaesthesiol Intensive Care Med 1:382
18. Cass NM, Lampard DG, Brown WA et al (1976) Computer controlled muscle relaxation: a comparison of four muscle relaxants in the sheep. Anaesthesiol Intensive Care Med 4:16–22
19. Brown TCK, Lampard DG, Brown WA et al (1976) Computer controlled studies in surface cooling hypothermia: the effect of carbon dioxide on brain temperature. Anaesth Intensive Care Med 4:36–40
20. Fleming RA, Smith NTy (1979) An inexpensive device for analyzing and monitoring the electroencephalogram. Anesthesiology 50:456–460

21. Fleming RA, Smith NTy (1979) Density modulation: a technique for the display of three-variable data in patient monitoring. Anesthesiology 50:543–546
22. Fleming RA, Coles JR, Smith NTy (1976) An inexpensive general purpose biomedical computer system. Proc San Diego Biomed Symp 15:79–89
23. Fleming RA, Coles JR, Smith NTy (1976) Patient protection of a biomedical computer system. Proc San Diego Biomed Symp 15:67–74

Regulation of the Depth of Anaesthesia: The Ultimate Goal of Quantitative Anaesthesia

H. Schwilden, H. Stoeckel, and J. Schüttler

The control and/or regulation of the depth of anaesthesia is predicated on the existence of a measurable pharmacodynamic signal as a correlate of depth of anaesthesia, as well as on the existence of a quantitative description of the influence of anaesthetics on this correlate.

Once an appropriate, measurable correlate of the depth of anaesthesia has been found, the identification of a model and the regulation of anaesthetic dosaging based on the model are primarily scientific problems, which, as was described in Dr. Westenkow's talk, can make use of the manifold developments in control and monitoring theory. The main challenge in the development of this field is in constructing a representation of the only too imprecisely defined clinical concept of depth of anaesthesia. The first efforts in this direction were made by Guedel [5], who published a paper in 1920 in which he used autonomic and neurological signs to define the stages of ether anaesthesia. However, even 60 years later, the idea that depth of anaesthesia is a graded phenomenon is not universally accepted, as demonstrated by an editorial in the *British Journal of Anaesthesia* [12].

Independent of these rather philosophic and semantic considerations, the goal of our present communication is to demonstrate that

1) The EEG can serve as the starting point for a measurement scale, the values of which correlate not only with the blood levels of the anaesthetics used, but also with the clinical signs of the state of consciousness or degree of vigilance;
2) The model, which describes a relation between dose and effect, can be used in the automatic regulation of the anaesthetic dosaging.

EEG Processing

The digital processing of the EEG requires the segmentation of the continuous signal into epochs of specified length. The choice of appropriate values for the epoch length T and digital sampling rate f depends on the characteristics of the signal. In order to conserve storage capacity, one tries, as a rule, to use the smallest possible frequency f. The Nyquist theorem sets the lower limit for f at two times the highest frequency contained in the signal. As EEG-frequency components higher than 30–40 Hz are generally only present in extremely small amounts, a sampling rate of 60–80 Hz would appear adequate. In practice, however, inductive pickup of power-line frequency ("hum") by the EEG leads cannot be ruled out. Thus, the Nyquist frequency must be greater than 120 Hz (line frequencies are typically 50 or 60 Hz). In our institute, a sapling rate of 125 Hz is used.

The longer the epoch, the wider the band pass of the system, but the poorer the temporal resolution of EEG trends (for changes in the EEG). We use the criterion that the wavelength of the lowest frequency must be shorter than 1/3 of the epoch duration. If a high-pass filter with a corner frequency of 0.5 Hz is used, an epoch duration greater than 6 s is necessary. Since our FFT-algorithm requires that the sample length (in ms) be an integer power of 2, an epoch duration of 8.192 s results. When Fourier transformations, which decompose the signal into amplitude and phase functions of frequency, are applied to the EEG with this epoch length, the phase relationships among the individual frequencies appear to be random (Fig. 1). Under these conditions, the power spectrum can be viewed as the actual information carrier.

Anaesthetics influence the power spectrum in characteristic ways. The waking state is frequently characterized by a peak in the region between 8–13 Hz (alpha-band). The effect of an anaesthetic is to push this peak towards lower frequencies. This may be seen in Fig. 2, which shows the effect of anaesthesia with enflurane and nitrous oxide. If the power spectrum is taken as a probability density function, the frequency shift may be parameterized using the methods of descriptive statistics. In general, the power spectrum is multimodal and skewed, so that the median (=50% quantile) [16–18, 20, 23] is an appropriate nonpara-

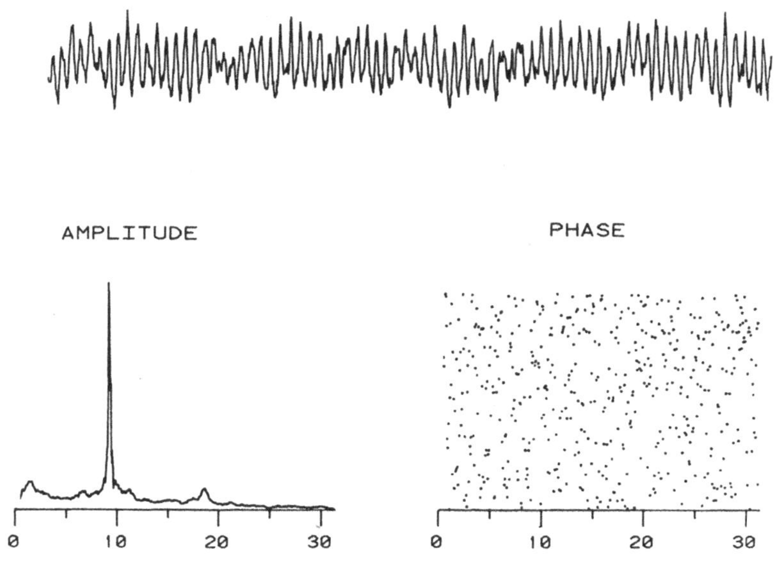

Fig. 1. The Fourier transformation depicts the amplitude and phase spectrum of the EEG epoch. The derived phase spectrum has the characteristics of statistical noise, and may thus be disregarded. The power spectrum is the square of the amplitude spectrum

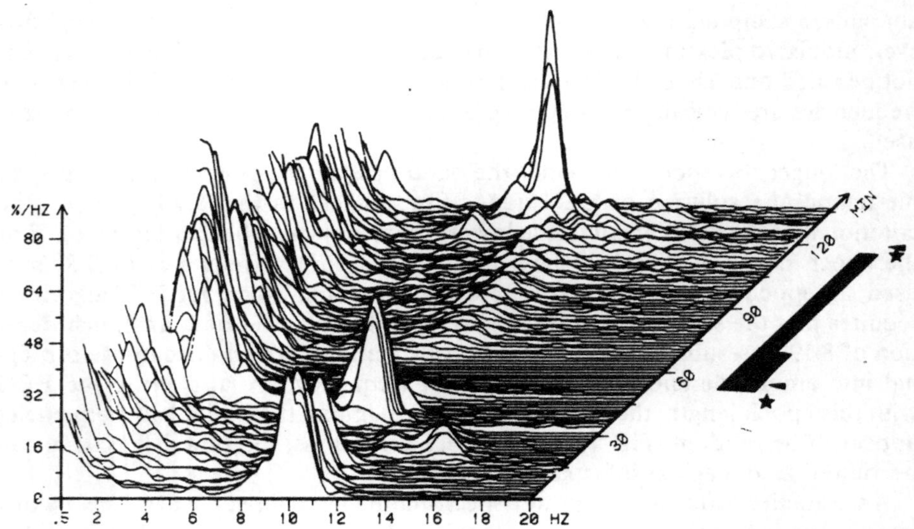

Fig. 2. Typical power spectrum taken during anaesthesia with enflurane in nitrous oxide

metric measure of the center of gravity of the distribution. In addition to this measure, the so-called edge frequency of the power spectrum (=95% quantile) is used, particularly in the Anglo-American literature [1, 8, 13, 21]. In anaesthesia, both measures are typically shifted toward lower frequencies.

Correlation of the EEG and Clinical Signs

Given the clinical observation that the median value falls during anaesthesia, we conducted several control studies in order to determine the range over which the median varies. In one class of experiments [10, 14, 18], linearly increasing blood levels of intravenous anaesthetics were used in order to determine, in a controlled manner, the order of the onset of anaesthetic effects.

Linear rises in the blood levels of the various anaesthetics were achieved with the aid of pharmacokinetic models for the individual substances and newer methods of intravenous dosing strategies [15, 19]. The concentration was raised until the burst suppression pattern appeared in the EEG. After terminating the infusion, we waited until the subject regained orientation to time and place. This cycle was then repeated twice. The intravenous hypnotics etomidate, methohexital, and propofol were investigated using this protocol. Figure 3 shows the temporal course of the power spectrum and the median for a subject in the propofol study.

Table 1 gives an overview of the median values for the 3 anaesthetics studied. The means and standard deviations of the medians are given for each of 4 stages defined by clinical signs. If the median is to be used as a variable for monitoring

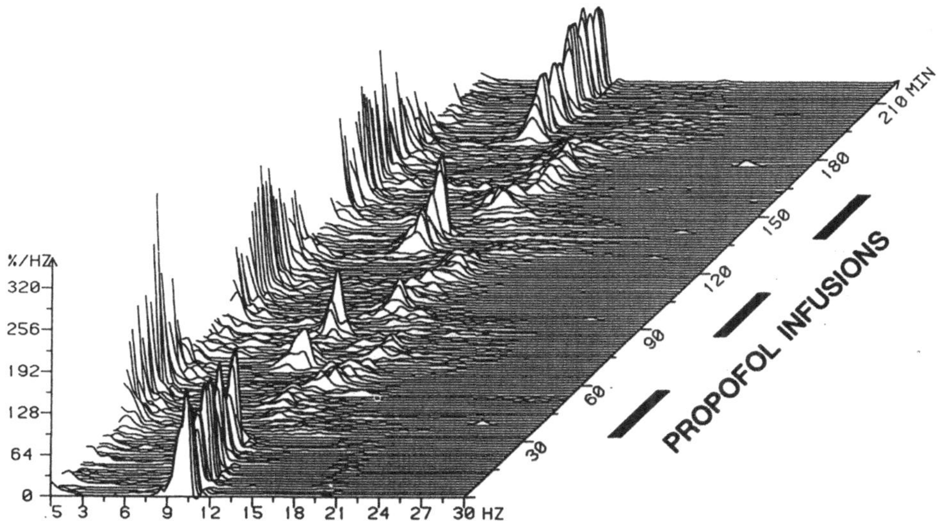

Fig. 3. *Above,* The power spectrum of a subject during the course of 3 intravenous applications of propofol given in order to achieve linearly increasing blood levels of the drug. *Below,* The course of the median value derived from the power spectrum

Table 1. Average values (±S.D.) of the medians in different clinically defined stages of anaesthesia

	Etomidate ($n=6$)	Methohexital ($n=6$)	Propofol ($n=7$)
A	9.45	9.60	9.76
	0.35	0.70	0.90
B	3.17	3.76	5.19
	0.42	1.02	2.15
C	2.83	2.69	2.85
	0.62	1.14	1.87
D	1.55	1.30	1.60
	0.31	0.40	0.16

A: preanaesthetic value; B: loss of lid reflex or lack of reaction to being spoken to; C: loss of corneal reflex; D: appearance of burst-suppression pattern

the adequacy of the depth of anaesthesia, then the range of values corresponding to an efficient hypnotic effect must be determined.

If one takes the view that the patient should not be conscious during the operation, then all median values must be ruled out, for which vigilant consciousness or the ability to react to being spoken to is present. This establishes a criterion for the upper limit of the median. A lower limit can not be so simply established, as the side effects of the anaesthetic on the circulation of the patient, as well as the desirability of a short postoperative recovery phase limits the dosaging of the agent. However, the burst suppression pattern is rarely seen in the clinical practice of anaesthesia. Although we can not furnish proof of it here, clinical experience would seem to indicate that burst suppression characterizes an unnecessarily deep level of anaesthesia, and thus serves to establish this criterion as the lower limit for the definition of a reasonable range of values for the median. Figure 4 shows cumulative distribution curves of the median frequency for all clinical experiments. The curve on the right gives the probability of occurrence of either a reaction to calling or a vigilance state, that on the left, the probability of the occurrence or reoccurrence of burst suppression. If one allows a 5% overlap of the distributions into the defined range, an interval from 2–4 Hz results. Under the stated premises then, this interval represents the optimal control range for the median of the power spectrum when levels of anaesthesia which are too light or too deep are to be avoided.

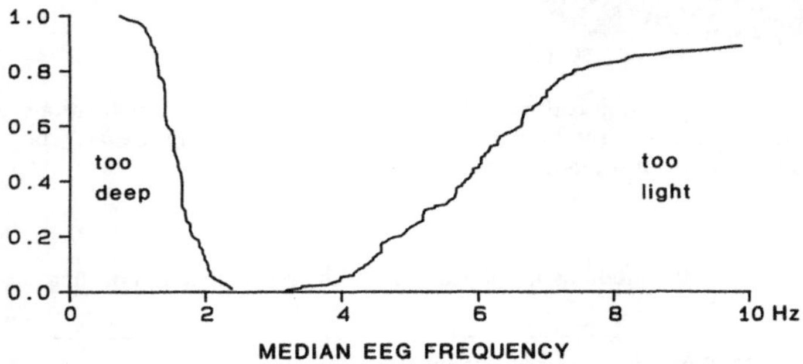

Fig. 4. Cumulative distributions of EEG median values. The distribution on the *right* pertains to values recorded when clinical signs indicated that the anaesthesia had become too light. The distribution on the *left*, to the occurrence of events (burst-suppression) indicating that the anaesthesia had become too deep. The probability of occurrence of a sign that the anaesthesia was too deep or too light is at a minimum for median values between 2 and 4 Hz

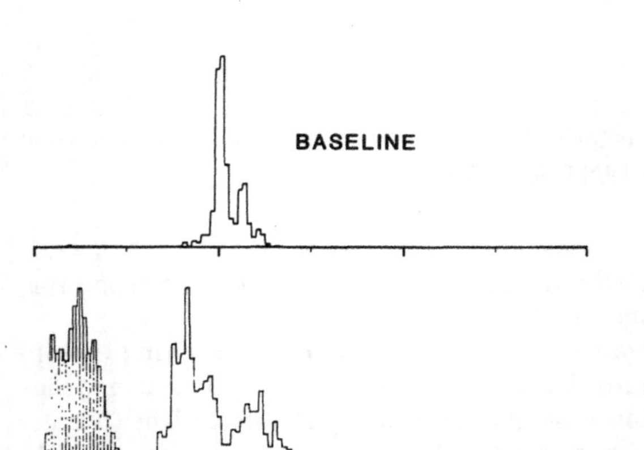

Fig. 5. Distribution of the medians of the power spectra recorded for a group of 14 patients anaesthetized with isoflurane in nitrous oxide. The preanaesthetic distribution is shown *above*, the distributions for the anaesthesia and recovery phases are given *below*

The 3 inhalation anaesthetics, halothane, enflurane, and isoflurane, were investigated in a controlled clinical study in which the anaesthetic was delivered in 60% nitrous oxide in end-expiratory concentrations of 1.3 and 1.5 MAC. Figure 5 shows the distribution of the median and edge frequencies in the waking state, during anaesthesia, and during the recovery phase after reaction to calling had reappeared for the isoflurane group [17]. One sees clearly that the distribution has an upper limit of 5 Hz during anaesthesia and a lower limit of 6 Hz during the recovery phase.

In Fig. 6, the relative activity of the EEG in various frequency bands is compared for the 3 anaesthetics and the waking state.

Taken together, the present results strongly indicate that a median value smaller than 4–5 Hz should be aimed at if intraoperative consciousness is to be avoided.

A qualification must be made. The median, as a single measure, is naturally inadequate as a representation of the complexity of a power spectrum which may be polymodal. For this reason, in pharmaco-EEG studies [4, 6, 7], which attempt to differentiate among the effects of various medications, a whole set of parameters is generally used to describe precisely the changes in the EEG. However, in order to keep within practicable limits, the goal of the present experiments was to identify EEG changes relevant to anaesthesia which could be characterized with the fewest possible parameters and to find one parameter which is as independent as possible of morphological changes in the EEG which are spe-

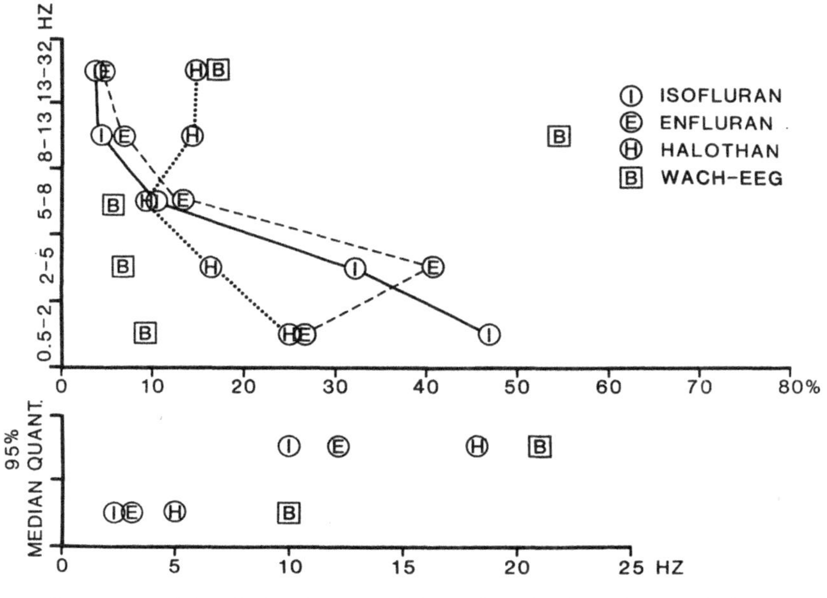

Fig. 6. The median values of the distributions of relative activity in the various EEG frequency bands and the median and edge (95% median quantile) frequencies are contrasted for three groups of 14 patients prior to and during anaesthesia. Each group received a different anaesthetic in 60% nitrous oxide: isoflurane, enflurane, or halothane

cific to a given medication. The ability to differentiate more finely among the effects on the electroencephalogram of the various anaesthetic agents and methods was sacrificed to this end. The question remains open, as to whether the present results regarding the state of anaesthesia are generally valid or not. The method used here, the correlation of EEG findings with clinical signs, is a purely empirical-phenomenological one. Thus, in the future, each individual medication will have to be tested as to whether the EEG correlates with the clinical signs, and as to whether the parameter values recommended here are also valid for each new agent.

Regulation of Anaesthetic Delivery According to the Electroencephalogram

Control loops are widely found in biological systems where a set point must be maintained or adjusted. Physiology has long been concerned with discovering how the control loops function which are necessary for the homeostasis of the animal and human organism. Control loops have been systematically applied in the natural and engineering sciences since the theory of control and feedback loops was developed in the area of cybernetics. There are 2 areas within pharmacological therapeutics in which the automatic dosing of an agent has found practical application beyond the experimental stage. These are the automatic regulation of blood sugar level with insulin and the regulation of blood pressure with the help of sodium nitroprusside.

Engineers differentiate between open-loop and closed-loop control [3, 24, 25]. Both methods require a model in order to determine the input necessary for a given output. In open-loop control, the input is independent of the previous output, whereas in closed-loop control, the input is partly determined by the previous output. Classical pharmacological therapy cannot be identified simply with one or the other of these 2 types of control. Prophylactic high-dose antibiotic therapy can be relatively easily identified with open-loop control, and the titrated dosing of sodium nitroprusside for the controlled lowering of blood pressure corresponds more or less to closed-loop control. In the latter case, the physician serves more than one function. He represents not only the superposed monitor in determining the set point of the system, but also an active element of the feedback loop in determining the dosaging of the pharmacological agent. In this case, determination of the dosaging rate is not based on a formalized model, but rather on the experience of the physician and the recommendations in the literature. The motivation behind the idea of automatic closed-loop control of drug dosaging is that by optimizing algorithms which determine the application rates according to the individual reactions of the patient, more space is created for the physician to make his therapeutic decisions (as the superposed monitor).

Control Algorithms

On-Off Control

The simplest form for a control loop is the "on-off" method. Whenever a particular effect occurs, the infusion of the drug is turned on (or off). The commonly practiced method of repetitive injection of a fixed quantity of an intravenous anaesthetic "according to need"can be seen as an example of on-off control. For instance, each time the blood pressure rises above a set threshold value, a fixed quantity of an analgesic is injected in a single bolus. In practice, the efficaciousness of this method is decidedly dependent on the pharmacokinetic behavior of the substance used.

P.I.D. Control

A commonly used control algorithm is the so-called proportional-integral-differential (P.I.D.) control. The rate, $\mathrm{I}(t)$, at which the gent is to be applied is determined by the sum of 3 terms. One term is directly proportional to the signal, a second is proportional to the (time) integral of the signal, and the third term is a function of the change of the signal with time:

$$\mathrm{I}(t) = k_1 \mathrm{E}(t) + k_2 \int_0^t \mathrm{d}t' \mathrm{E}(t') + k_3 \frac{\mathrm{d}E}{\mathrm{d}t} + I_0$$

The proportional term is a function of the momentary measurement, whereas the integral and differential terms reflect global and local temporal trends. The choice of the constants k_i, $i = 1, 2, 3$ are critical to the performance of this type of model. Generally, the values of these constants are determined in prior control experiments and their values significantly affect the quality of the control loop behavior. Obviously, P.I.D. control is primarily suited for use with linear systems. Use with nonlinear systems, as is the case in the present study, requires restricting application to approximately linear portions, as has been suggested by Asbury [2] with respect to the control of muscle relaxation.

Both the on-off and the P.I.D. control systems represent heuristic methods in that they require no knowledge of cause-effect relations. If the feedback signal is used not only for determining te input function, but also for correcting the parameters of the control algorithm (e.g., the values k_i, $i = 1, 2, 3$ for P.I.D. control), then one speaks of adaptive control.

Input-Output Control Model

In a recently published article [20], a control loop was constructed and experimentally tested, which made use of the information provided by a pharmacological input-output model in order to determine the input function of the pharmacological agent. The median of the EEG-power spectrum served as the feedback signal. Design of the control loop was guided by knowledge of the pharmacokinetic and pharmacodynamic model.

The pharmacodynamic model allows the determination of the concentration of the drug necessary for a desired effect. The pharmacokinetic model allows the determination of the dosaging function necessary to achieve and maintain the desired concentration in the fastest possible manner.

The following relations among the dosaging function $I(t)$, the concentration $c(t)$, and the effect $E(t)$ were assumed:

$$c(t) = \int_0^t dt' G(t-t') I(t) \tag{1}$$

where $G(t) = A\exp(-\alpha t) + B\exp(-\beta t)$ and

$$E(t) = E_1 - E_2 \frac{c^{\gamma}(t)}{c_0^{\gamma} + c^{\gamma}(t)} \tag{2}$$

If $E(t)$ is to remain constant over time as E_s, then a time stationary concentration c_s may be substituted into Eq. (1):

$$c_s = c_o [(E_1 - E_s)/(E_2 - E_1 + E_s)]^{1/\delta}$$

Rearranging Eq. (1),

$$I(t) = \int_0^t dt' G^{-1}(t-t') c(t')$$

If $G(t)$ is represented as the bi-exponential equation:

$$G(t) = A\exp(-\alpha t) + B\exp(-\beta t),$$

then the inverse, G^{-1}, is

$$G^{-1}(t) = \delta'(t) + [k_{el} + k_{12}\exp(-k_{21}t)]\Theta(t), \tag{3}$$

whereby k_{el}, k_{12}, and k_{21} are given by

$$k_{el} = (A+B)/(A/\alpha + B/\beta)$$
$$k_{21} = \alpha\beta/k_{el}$$
$$k_{12} = \alpha + \beta - k_{el} - k_{21}.$$

The formula is still not usable for control purposes. In general, a given dosing strategy must be corrected at some time t_1, when the signal monitor determines a deviation from the set point.

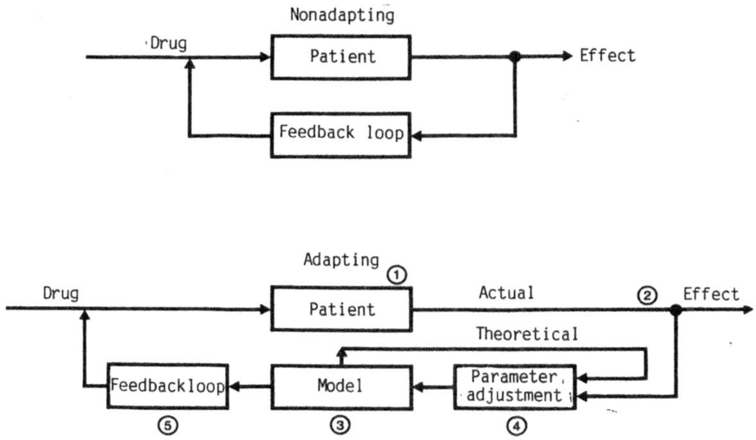

Fig. 7. Comparison of a nonadapting and an adapting control loop. The process which is to be regulated *(1)* is represented by a signal *(2)*. This signal is compared with the results of model calculations *(3)*. In the case of noncorrespondence, parameters are adjusted such as to bring the model calculations into line with the measured signal *(4)*. The controller *(5)* then determines the dosaging, based on the model

If E_i is the median value measured at time t_i, then Eq. (1) gives the concentration c_i, which, by an appropriate dosing measure, must be corrected to c_s. Two cases must be distinguished here: $c_i > c_s$ and $c_i < c_s$.

In general, if $c_i > c_s$, dosaging can only be cut off until the concentration has fallen so far that the set point is reached again. In the second case, one must consider the amounts of the drug M_1 and M_2 contained in the central and peripheral distribution spaces. Equation 3 must then be generalized as follows, taking on the following explicit form:

$$I(t) = \delta(t)(M_s - M_1) + M_s k_{el} + (k_{12} M_s - k_{21} M_2) e^{-k_{21} t},$$

whereby M_s is the amount of the drug corresponding to concentration c_s in the initial distribution space. In order to adjust the input-output model simultaneously, it would be ideal to determine all parameters of the model during the regulation period. Furthermore, the greatest possible signal variation would be required to achieve a dependable estimation of the parameters, as was done in the control studies. However, the goal of the regulation is to keep variation in the signal as small as possible. It is thus practical to adjust as few parameters as possible. The parameters A and B were chosen for adjustment in the present work. This set of parameters, with which the two pharmacokinetic variables representing initial distribution space and clearance can be varied, allows the adjustment of the short-term and long-term values determining dosaging.

In detail, the effect E is expressed as a function of the 2 parameters A and B.

$$E = E(A, B)$$

If the actual hybrid constants of the subject are written as $A+\delta A$ and $B+\delta B$, then the difference (E) between the measured signal and the model prediction can be written as a Taylor series:

$$\Delta E = E(A+\delta A, B+\delta B) - E(A,B)$$
$$= (\partial E/\partial A\,\delta A + \partial E/\partial B\,\delta B) + \cdots$$

This equation was used to determine the new hybrid constants by minimizing the expression $\delta A^2 + \delta B^2$.

The automatic closed-loop regulation of methohexital anaesthesia via the electroencephalogram was tested on 13-min resting phase, the base line EEG was recorded while the subjects listened to music over headphones. The music was used to minimize excitation and to mask the onset of methohexital infusion. The goal of the automatic regulation was to hold the median of the EEG power spectrum between 2 and 3 Hz for 2 h. As soon as the median value came within the control interval, the subject received a random stimulus approximately once every 1.5 min. Stimuli included cold stimuli, light pin pricks, noises, and verbal commands, as well as testing of the lid reflex and corneal reflex.

The choice of these stimuli was based on the experience gained with the first 2 subjects who were treated somewhat differently. The first subject was not stimulated at all. In the 1st h, the second subject received electrical stimulation every 20 s. This stimulus consisted of a supramaximal train of 4 pulses to the Nervus medianus, as recommended for the monitoring of muscle relaxation during anaesthesia. In the 2nd h of the automatically controlled anaesthesia, this subject received the random stimuli of varying quality as described above.

With the first, unstimulated subject, the control loop went into oscillation. Figure 8 shows the temporal course of the median, edge frequency, and blood levels of methohexital which were calculated on the basis of average pharmacokinetic

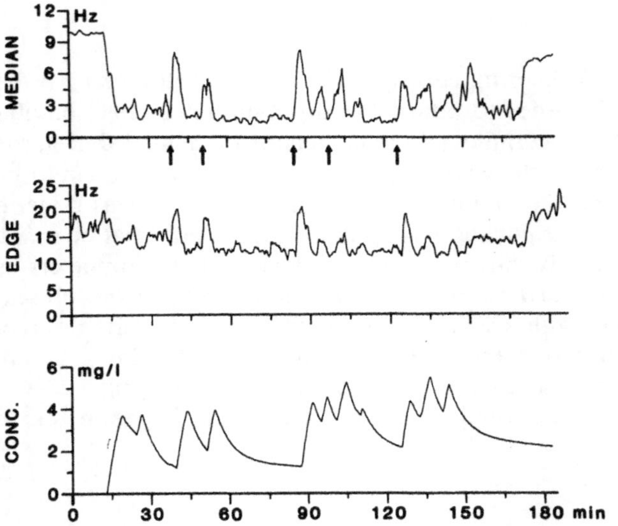

Fig. 8. The time course of the median, edge frequency, and calculated concentration of methohexital in the blood for the first subject in the control study. The *arrows* denote waking episodes

data. Clinically, this subject showed clear waking reactions five times. These episodes are marked with arrows. The control loop reacted correctly each time, bringing the median immediately back within the control interval.

A possible cause of these oscillations may be seen in the course of the calculated blood-level values. In four of the five cases in which waking reactions occurred, a clear fall in the blood level preceded the episode.

During the 1st h of closed-loop control, the second subject received supramaximal electrical stimulation of the nervus medianus, consisting of a train of four pulses every 20 s. After approximately 1 h, this subject also showed a waking reaction preceded by a longer interval of falling methohexital blood levels. For this reason, the random stimulation protocol described above was introduced 60 min after the beginning of the automatically controlled methohexital anaesthesia. No further waking reactions were observed during the rest of the anaesthesia, and the control loop remained stabile.

The random stimulus protocol was used over the entire duration of the automated anaesthesia with 11 further subjects. The average time (±S.D.) required to reach the control interval of 2–3 Hz was 9.6±5.2 min. During this time, 192±103 mg methohexital were infused.

The average values of all EEG parameters obtained during the control period are given in Table 2. In accordance with the target interval for the median of

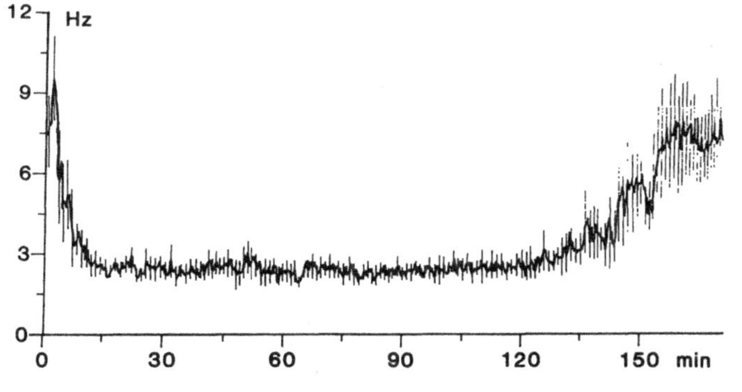

Fig. 9. Time course of the average (±S.D.) values of the medians of the EEG power spectra of 11 subjects during an automated anaesthesia with methohexital. The control loop was designed to keep the median value between 2 and 3 Hz

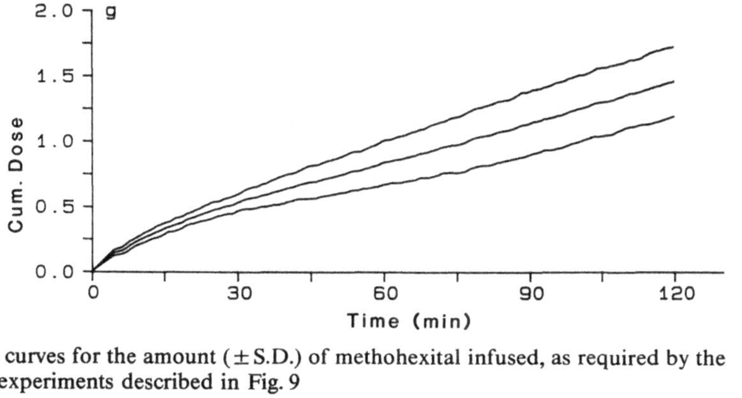

Fig. 10. Cumulative curves for the amount (±S.D.) of methohexital infused, as required by the control loop in the experiments described in Fig. 9

Table 2. Relative EEG activity, median, edge, and amplitude during the waking (preanaesthetic) phase (B) and during automated anaesthesia for 11 subjects

	Relative EEG activity (%)										Median		Edge		Amplitude	
Subject	0.5–2 Hz		>2–3 Hz		>5–8 Hz		>8–13 Hz		>13–32 Hz		(Hz)		(Hz)		(μV)	
	B	F	B	F	B	F	B	F	B	F	B	F	B	F	B	F
3	13.6	46.9	10.3	23.8	13.4	8.7	36.3	14.5	26.2	6.1	9.1	2.4	21.2	14.4	14.4	41.5
4	12.7	45.2	13.2	19.1	15.0	8.0	37.3	24.1	21.8	3.6	8.9	2.7	22.1	11.8	21.2	77.3
5	10.7	42.3	9.1	28.0	5.3	8.1	68.7	16.0	6.2	5.5	10.9	2.5	13.4	13.1	12.4	23.0
6	5.6	46.1	6.1	17.4	8.4	9.1	72.4	19.7	7.5	7.7	8.9	2.8	15.9	14.8	8.6	14.9
7	17.7	46.4	10.5	25.5	11.1	8.9	31.2	13.9	29.5	5.3	9.4	2.3	26.1	12.8	6.9	16.6
8	5.2	47.8	7.2	17.4	5.0	8.2	70.0	19.8	12.5	6.9	10.8	2.5	20.2	14.0	20.2	56.8
9	16.9	43.8	21.7	21.9	12.2	8.3	13.5	18.3	35.7	7.6	8.5	2.6	23.5	14.5	11.8	74.3
10	15.0	49.2	10.7	20.0	18.1	10.3	34.1	13.7	22.1	6.8	8.6	2.4	21.6	13.9	6.8	17.0
11	10.9	48.0	10.0	22.3	17.2	11.6	51.4	11.1	10.5	6.9	8.7	2.4	17.1	13.7	21.3	48.8
12	15.2	42.3	10.5	28.3	16.1	9.4	27.0	14.4	31.2	5.6	9.1	2.6	22.8	13.0	10.7	13.9
13	16.3	48.4	19.3	20.6	8.6	9.0	28.4	15.9	27.3	6.0	9.2	2.4	20.1	13.4	34.4	36.6
Median	12.5	46.0	11.8	22.1	11.7	9.1	43.4	16.9	20.4	6.1	9.3	2.5	20.3	13.5	15.4	37.9
±SD	4.5	2.5	5.0	4.1	4.8	1.1	20.1	3.8	10.6	1.2	0.9	0.2	3.8	0.9	8.8	24.8

2–3 Hz, the activity in the δ_1-band was just under 50%. The time course of the average value ($\pm$S.D.) of the median is shown in Fig. 9. The cumulative methohexital dosage (average$\pm$S.D.) applied during these experiments, as required by the control loop, is shown in Fig. 10.

References

1. Arden JR, Holley FO, Stanski DR (1986) Increased sensitivity to etomidate in the elderly: initial distribution versus altered brain response. Anesthesiology 65:19
2. Asbury AJ (1985) Feedback control and model identification in the automatic control of neuromuscular blockade. In: Stoeckel H (ed) Quantitation, modelling and control in anaesthesia. Thieme, Stuttgart, p 274
3. Cosgrove RJ: Automatic, feedback controlled drug delivery devices. In: Smolen F, Bull LA (ed) Controlled drug bioavailability, vol 3. Wiley, New York, p 109
4. Fink M (1977) Quantitative EEG analysis and psychopharmacoloty. In: Remond A (ed) EEG informatics. A didactic review of methods and application of EEG data processing. Elsevier/North Holland, Amsterdam, p 301
5. Guedel, AE (1920) Third stage ether anesthesia: a subclassification regarding the significance of the position and movement of the eyeball. Am J Surg Anesth (Suppl) 34:53
6. Herrmann WM (1982) Development and critical evaluation of an objective procedure for the electroencephalographic classification of psychotropic drugs. In: Herrmann WM (ed) Electroencephaolography in drug research. Fischer, Stuttgart, p 249
7. Herrmann WM (ed) (1982) Electroencephalography in drug research. Fischer, Stuttgart
8. Hudson RJ, Stanski DR, Saidman LJ, Meathe E (1983) A model for studying depth of anesthesia and acute tolerance to thiopentone. Anesthesiology 59:301
9. Inoue M, Kajiya F, Inada H, Kitabatake A et al (1976) Optimal control of medical treatment: adaptive control of blood glucose level in diabetic coma. Comput Biomed Res 9:217
10. Lauven PM, Stoeckel H, Schwilden H (1985) Venous threshold concentrations of methohexitone (abstract). Anesthesiology 63 (Suppl):A368
11. Low JM, Millard RK, Curnow JSH, Hutton P, Prys-Roberts C (1985) Servo-controlled infusion of sodium nitroprusside for induced hypotension. In: Stoeckel H (ed) Quantitation, modelling and control in anaesthesia. Thieme, Stuttgart, p 286
12. Prys-Roberts C (1987) Anaesthesia: a practical or impractical construct? Br J Anaesth 59:1341
13. Rampil IJ, Sasse FJ, Smith NT, Hoff BH, Fleming DC (1980) Spectral edge frequency – a new correlate of anesthetic depth. Anesthesiology 53:S12
14. Schüttler J, Schwilden H, Stoeckel H (1985) Infusion strategies to investigate the pharmacokinetics and pharmacodynamics of hypnotic drugs: etomidate as an example. Eur J Anaesthesiol 2:133
15. Schwilden H (1986) Die interaktive Dosierung intravenöser Anästhetika. Thieme, Stuttgart
16. Schwilden H, Stoeckel H (1980) Untersuchungen über verschiedene EEG-Parameter als Indikatoren des Narkosezustandes. Der Median als quantitatives Maß der Narkosetiefe. Anästh Intensivther Notfallmed 15:279
17. Schwilden H, Stoeckel H (1987) Quantitative EEG analysis during anaesthesia with isoflurane in nitrous oxide at 1.3 and 1.5 MAC. Br J Anaesth 59:738
18. Schwilden H, Schüttler J, Stoeckel H (1985) Quantitation of the EEG and pharmacodynamic modelling of hypnotic drugs: etomidate as an example. Eur J Anaesthesiol 2:121
19. Schwilden H, Stoeckel H, Schüttler J, Lauven PM (1986) Pharmacological models and their use in clinical anaesthesia. Eur J Anaesthesiol 3:175
20. Schwilden H, Schüttler J, Stoeckel H (1987) Closed-loop feedback control of methohexital anesthesia by quantitative EEG analysis in humans. Anesthesiology 67:341

21. Scott JC, Ponganis KV, Stanski DR (1985) EEG quantitation of narcotic effect: the comparative pharmacodynamics of fentanyl and alfentanil. Anesthesiology 62:234
22. Sheppart LC (1981) Computer control of the infusion of vasoactive drugs. Ann Biomed Eng 8:431
23. Stoeckel H, Schwilden H (1984) Quantitative EEG-analysis and monitoring depth of anaesthesia. In: Gomez QJ, Egay LM, de la Cruz-Odi MF (eds) Anesthesia - safety for all. Elsevier, Amsterdam, p 151
24. Stoeckel H, Schwilden H (1986) Methoden der automatischen Feedback-Regelung für die Narkose. Konzepte und klinische Anwendung. Anästh Intensivmed Notfallmed 21:60
25. Vozeh S, Steimer J-L (1985) Feedback control methods for drug dosage optimisation. Concepts, classification and clinical application. Clin Pharmacokinet 10:457
26. Westenskow DR, Meline LJ, Pace NL, Bodily MN (1984) Sodium nitroprusside induced hypotension with computer adjustment for varying drug sensitivity. Anesth Analg 63:281

GPSR Compliance

The European Union's (EU) General Product Safety Regulation (GPSR) is a set of rules that requires consumer products to be safe and our obligations to ensure this.

If you have any concerns about our products, you can contact us on ProductSafety@springernature.com

In case Publisher is established outside the EU, the EU authorized representative is:

Springer Nature Customer Service Center GmbH
Europaplatz 3
69115 Heidelberg, Germany

Batch number: 10704771

Printed by Printforce, the Netherlands